P9-ASA-648

A. ŠVEC · GLOBAL DIFFERENTIAL GEOMETRY OF SURFACES

Global Differential Geometry of Surfaces

BY A. ŠVEC

D. Reidel Publishing Company

Dordrecht: Holland / Boston: U.S.A. / London: England

Library of Congress Cataloging in Publication Data

CIP

Švec, Alois.
Global differential geometry of surfaces.

Bibliography: p.
1. Geometry, Differential. 2. Surfaces. I. Title.
QA641.S86 1981 516.3′6 81-7334
ISBN 90-277-1295-6 AACR2

Distributors for the U.S.A. and Canada
Kluwer Boston Inc.
190 Old Derby Street, Hingham, MA 02043, U.S.A.

Distributors for all remaining countries
D. Reidel Publishing Company,
P.O. Box 17, 3300 AA Dordrecht, Holland

D. Reidel Publishing Company is a member of the Kluwer Group.

Printed in the German Democratic Republic.

Preface

Writing this book, I had in my mind a reader trying to get some knowledge of a part of the modern differential geometry. I concentrate myself on the study of surfaces in the Euclidean 3-space, this being the most natural object for investigation. The global differential geometry of surfaces in E^3 is based on two classical results: (i) the ovaloids (i.e., closed surfaces with positive Gauss curvature) with constant Gauss or mean curvature are the spheres, (ii) two isometric ovaloids are congruent. The results presented here show vast generalizations of these facts.

Up to now, there is only one book covering this area of research: the Lecture Notes [3] written in the tensor slang. In my book, I am using the machinary of E. Cartan's calculus. It should be equivalent to the tensor calculus; nevertheless, using it I get better results (but, honestly, sometimes it is too complicated). It may be said that almost all results are new and belong to myself (the exceptions being the introductory three chapters, the few classical results and results of my post-graduate student Mr. M. Afwat who proved Theorems V.3.1, V.3.3 and VIII.2.1—6).

The first three chapters are quite standard and just make the book complete; the reader may very well consult other sources. The fourth chapter contains the necessary parts of the local geometry and preliminary calculations. The fifth and sixth chapters constitute the main parts of the book. All results are new (i.e., they are proper generalizations of the known results). The last two chapters serve as an appendix showing two possibilities of further studies by means of the previously used methods. The bibliography is limited just to reference concerning the maximum principle [2] and the pseudoanalytic functions [4]; in [1], we present further generalizations of the material contained in the last chapter.

Praha, Summer 1978

Alois Švec

Contents

I. Multilinear algebra

1. Vector spaces

Let $\mathbf{R}$ denote the field of real numbers.

Definition 1.1. *A* vector space V *over* $\mathbf{R}$ *is a set* V *(of the so-called* vectors) *together with two mappings*

$$(1.1)\qquad V \times V \to V,\quad (v_1, v_2) \mapsto v_1 + v_2;\qquad \mathbf{R} \times V \to V,\quad (r, v) \mapsto rv,$$

satisfying the following axioms:

(i) $v + v_1 = v_1 + v$;

(ii) $(v + v_1) + v_2 = v + (v_1 + v_2)$;

(iii) *there is a vector* $0 \in V$ *such that* $0 + v = v$;

(iv) *for each* $v \in V$, *there is a vector* $-v \in V$ *such that* $v + (-v) = 0$;

(v) $r(v + v_1) = rv + rv_1$;

(vi) $(r_1 + r_2)\, v = r_1 v + r_2 v$;

(vii) $(r_1 r_2)\, v = r_1(r_2 v)$;

(viii) $1 \cdot v = v$

for all $v, v_1, v_2 \in V$; $r, r_1, r_2 \in \mathbf{R}$. *A subset* $W \subset V$ *is called the* subspace *of* V *if* $w, w_1 \in W$; $r, r_1 \in \mathbf{R}$ *implies* $rw + r_1 w_1 \in W$. *Let* $M \subset V$; *denote by* $h(M)$ *the subspace of all vectors of the form*

$$(1.2)\qquad v = r_1 v_1 + \cdots + r_m v_m;\quad v_1, \ldots, v_m \in M;\quad r_1, \ldots, r_m \in \mathbf{R}.$$

The vectors $v_1, \ldots, v_m \in V$ *are said to be* linearly independent *if* $r_1 v_1 + \cdots + r_m v_m = 0$ *implies* $r_1 = \cdots = r_m = 0$. *The vectors* $v_1, \ldots, v_n$ *form a* basis *of* V *if they are linearly independent and* $h\{v_1, \ldots, v_n\} = V$. *We say that* V *is* finite-dimensional *if it has a finite basis* $v_1, \ldots, v_n$. *It is easy to see that any other basis of* V *has the same number* n *of elements called the* dimension *of* V.

Definition 1.2. *Let* V, W *be vector spaces and* $f\colon V \to W$ *a mapping. The mapping* f *is called a* homomorphism *if*

$$(1.3)\qquad f(r_1 v_1 + r_2 v_2) = r_1 f(v_1) + r_2 f(v_2)\quad \text{for all}\quad r_1, r_2 \in \mathbf{R};\quad v_1, v_2 \in V.$$

The image *and the* kernel *of the homomorphism* $f\colon V \to W$ *be defined as*

$$(1.4)\qquad \operatorname{Im} f = f(V) \subset W,\qquad \operatorname{Ker} f = f^{-1}(0) \subset V$$

resp.; obviously, $\operatorname{Im} f$ *and* $\operatorname{Ker} f$ *are subspaces. Consider the sequence*

$$(1.5)\qquad U \xrightarrow{f} V \xrightarrow{g} W,$$

U, V, W being vector spaces and f, g homomorphisms. The sequence (1.5) *is called* exact *if* $\operatorname{Im} f = \operatorname{Ker} g$. *Let* $\{0\}$ *denote the* trivial vector space *consisting just of the zero vector. The homomorphism* $f: V \to W$ *is called a* monomorphism *or an* epimorphism *if the sequence*

$$(1.6) \qquad \{0\} \to V \xrightarrow{f} W \quad \text{or} \quad V \xrightarrow{f} W \to \{0\}$$

resp. is exact; here, $\{0\} \to V$ *and* $W \to \{0\}$ *are the natural homomorphisms. An* isomorphism *is a homomorphism which is a monomorphism and, at the same time, an epimorphism.*

Let V, W be vector spaces and $L(V, W)$ the set of all homomorphisms $f: V \to W$. For $f_1, f_2 \in L(V, W)$, $r \in \mathbf{R}$, define $f_1 + f_2$, $rf_1 \in L(V, W)$ by the relation

$$(1.7) \qquad (f_1 + g_2)(v) = f_1(v) + f_2(v), \quad (rf_1)(v) = rf_1(v) \quad \text{for each} \quad v \in V.$$

With respect to these operations, $L(V, W)$ becomes a vector space, and we have

$$(1.8) \qquad \dim L(V, W) = \dim V \cdot \dim W.$$

Definition 1.3. *The space* $V^* = L(V, \mathbf{R})$ *is called the* dual space *of* V.

Introduce the so-called Kronecker deltas by

$$(1.9) \qquad \delta_i^j = 1 \quad \text{for} \quad i = j, \qquad \delta_i^j = 0 \quad \text{for} \quad i \neq j; \qquad i, j, \ldots = 1, 2, \ldots$$

Let $v_1, \ldots, v_n$ be a basis of V; then there is exactly one basis $v^1, \ldots, v^n$ of V^* (called the *dual basis*) such that $v^i(v_j) = \delta_j^i$ for $i, j = 1, \ldots, n$. Let us construct a mapping $\iota: V \to V^{**} = (V^*)^*$ as follows: for $v \in V$, define $\iota(v) \in L(V^*, \mathbf{R})$ by the relation $\iota(v)(f) = f(v)$ for each $f \in V^*$. Prove that ι is an isomorphism for a finite-dimensional V. (Attention: If $\dim V = \infty$, ι is just a monomorphism!)

Theorem 1.1. *Let* $g: V \to W$ *be a homomorphism. Then there is exactly one homomorphism* $g^*: W^* \to V^*$ *such that*

$$(1.10) \qquad g^*(w^*)(v) = w^*\big(g(v)\big) \quad \textit{for each} \quad w^* \in W^*, \ v \in V;$$

g^* *is called the* dual homomorphism. *The sequence* (1.5) *being exact, the sequence*

$$(1.11) \qquad W^* \xrightarrow{g^*} V^* \xrightarrow{f^*} U^*$$

is exact as well.

Proof. It is sufficient to prove that (1.11) is exact. For each vector space T, define the mapping

$$(1.12) \qquad \langle\ ,\ \rangle_T: T \times T^* \to \mathbf{R}, \qquad \langle t, t^* \rangle_T = t^*(t).$$

The equation (1.10) takes the form

$$(1.13) \qquad \langle v, f^*(w^*) \rangle_V = \langle f(v), w^* \rangle_W.$$

For $w^* \in W^*$, $u \in U$, we get

$$(1.14) \qquad \big\langle u, f^*\big(g^*(w^*)\big)\big\rangle_U = \langle f(u), g^*(w^*) \rangle_V = \big\langle g\big(f(u)\big), w^* \big\rangle_W;$$

(1.5) being exact, we have $g(f(u)) = 0$, i.e., $f^*(g^*(w^*)) = 0$ and $\operatorname{Im} g^* \subset \operatorname{Ker} f^*$. Now, let us prove $\operatorname{Ker} f^* \subset \operatorname{Im} g^*$. Let $v^* \in \operatorname{Ker} f^*$, i.e., $f^*(v^*) = 0$. For $u \in U$,

$$(1.15) \qquad 0 = \langle u, f^*(v^*)\rangle_U = \langle f(u), v^*\rangle_V .$$

The homomorphism $\tilde{w}^*: \operatorname{Im} g \to \mathbf{R}$ be defined as follows. For $w \in \operatorname{Im} g$, let us choose $v \in V$ such that $g(v) = w$ and define $\tilde{w}^*(w) = v^*(v)$. This is a good definition: $v_1 \in V$ satisfying $g(v_1) = w$, we have $v - v_1 \in \operatorname{Ker} g = \operatorname{Im} f$, i.e., there exists $u \in U$ such that $v - v_1 = f(u)$; from (1.15), $0 = \langle v - v_1, v^*\rangle_V$, i.e., $v^*(v) = v^*(v_1)$. Now, let us choose any homomorphism $w^*: W \to \mathbf{R}$ such that its restriction to $\operatorname{Im} g \subset W$ is equal to $\tilde{w}^*$. For any $v \in V$, $g(v) \in \operatorname{Im} g$, and the definition of w^* implies

$$\langle v, g^*(w^*)\rangle_V = \langle g(v), w^*\rangle_W = \langle v, v^*\rangle_V .$$

Thus $v^* = g^*(w^*)$, i.e., $v^* \in \operatorname{Im} g^*$. QED.

Let $U \subset V$ be vector spaces. Let V/U be the set of all subsets of the form $L_v = \{v + u; u \in U\}$, $v \in V$. Define $r_1 L_{v_1} + r_2 L_{v_2}$ as $L_{r_1 v_1 + r_2 v_2}$. It is easy to see that $L_{v_1'} = L_{v_1}$, $L_{v_2'} = L_{v_2}$ implies $L_{r_1 v_1 + r_2 v_2} = L_{r_1 v_1' + r_2 v_2'}$. Thus V/U gets the structure of a vector space, the so-called *factor-space*. Further, we get the natural homomorphism $V \to V/U$, $v \mapsto L_v$.

2. Tensor products

Definition 2.1. *Let $V_1, \ldots, V_r, W$ be vector spaces. The mapping*

$$(2.1) \qquad f: V_1 \times \cdots \times V_r \to W$$

is called multilinear *if*

$$(2.2) \qquad f(v_1, \ldots, s v_i + s' v_i', \ldots, v_r) = s f(v_1, \ldots, v_i, \ldots, v_r) + s' f(v_1, \ldots, v_i', \ldots, v_r)$$

for each $v_1 \in V_1; \ldots; v_i, v_i' \in V_i; \ldots; v_r \in V_r$; $s, s' \in \mathbf{R}$; $i = 1, \ldots, r$.

Denote by $L(V_1, \ldots, V_r; W)$ the set of all multilinear mappings (2.1); the set $L(V_1, \ldots, V_r; W)$ is made into a vector space by

$$(2.3) \qquad \begin{aligned} (f_1 + f_2)(v_1, \ldots, v_r) &= f_1(v_1, \ldots, v_r) + f_2(v_1, \ldots, v_r), \\ (sf)(v_1, \ldots, v_r) &= s f(v_1, \ldots, v_r). \end{aligned}$$

Definition 2.2. *The* tensor product *of the vector spaces U, V is any couple (T, τ), T being a vector space and $\tau \in L(U, V; T)$, such that:* (i) *the linear hull of $\tau(U \times V)$ is equal to T;* (ii) *W being an arbitrary vector space and $f \in L(U, V; W)$; there is a homomorphism $g: T \to W$ such that the diagram*

$$(2.4) \qquad \begin{array}{ccc} & U \times V & \\ {}^{\tau}\swarrow & & \searrow{}^{f} \\ T & \xrightarrow[g]{\quad - - - \quad} & W \end{array}$$

is commutative.

Theorem 2.1. *The homomorphism g of* (ii) Def. 2.2 *is unique.* (T, τ) *and* (T', τ') *being two tensor products of U and V, there is an isomorphism $\sigma: T \to T'$ such that the diagram*

(2.5)
$$\begin{array}{ccc} & U \times V & \\ \tau \swarrow & & \searrow \tau' \\ T & \xrightarrow[\sigma]{} & T' \end{array}$$

is commutative.

Proof. Let $g_1, g_2: T \to W$ be homomorphisms such that $f = g_1 \circ \tau = g_2 \circ \tau$. Then $g_1 - g_2$ maps $\tau(U \times V)$ into $0 \in W$, and, the linear hull of $\tau(U \times V)$ being T, $(g_1 - g_2)(T) = 0 \in W$. Thus $g_1 = g_2$. Because of (ii) Def. 2.2, there are homomorphisms σ, σ' such that the diagram

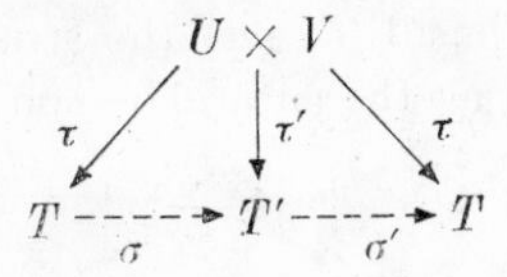

is commutative. Thus, $\sigma \circ \sigma'$ is the identity mapping on $\tau(U \times V) \subset T$, and, consequently, on T. Analogously, $\sigma' \circ \sigma = \mathrm{id}: T \to T$. QED.

Theorem 2.2. *There exists a tensor product of two vector spaces.*

Proof. We are going to produce a direct construction. Let Q be the set of all formal finite sums of the form

$$a_1(u_1, v_1) + \cdots + a_r(u_r, v_r); \quad a_i \in \mathbf{R}, \quad u_i \in U, \quad v_i \in V. \tag{2.6}$$

Obviously, Q is a vector space in a natural way. Q might be also interpreted as the set of functions $a: U \times V \to \mathbf{R}$ taking non-zero values on a finite number of points of $U \times V$. To the formal sum (2.6), we associate the function a such that $a\big((u_i, v_i)\big) = a_i$ for $i = 1, \ldots, r$ and $a(w) = 0$ for all other points $w \in U \times V$. Of course, $U \times V \subset Q$. Let $N' \subset Q$ be the set of all elements of the form

$$\begin{aligned} q = (a_1u_1 + a_2u_2, b_1v_1 + b_2v_2) - a_1b_1(u_1, v_1) - a_1b_2(u_1, v_2) - a_2b_1(u_2, v_1) \\ - a_2b_2(u_2, v_2); \\ a_1, a_2, b_1, b_2 \in \mathbf{R}; \quad u_1, u_2 \in U; \quad v_1, v_2 \in V. \end{aligned} \tag{2.7}$$

Let N be the linear hull of N', $U \otimes V = Q/N$, and let $\otimes: U \times V \to U \otimes V$ be the restriction of the natural homomorphism $\varkappa: Q \to Q/N$ to the subset $U \times V \subset Q$.

Let us prove that $(U \otimes V, \otimes)$ is the tensor product of U and V. Consider the point $q \in Q$ (2.7); then $\varkappa(q) = 0$, i.e.,

$$\begin{aligned} (a_1u_1 + a_2u_2) \otimes (b_1v_1 + b_2v_2) = a_1b_1 \cdot u_1 \otimes v_1 + a_1b_2 \cdot u_1 \otimes v_2 \\ + a_2b_1 \cdot u_2 \otimes v_1 + a_2b_2 \cdot u_2 \otimes v_2, \end{aligned} \tag{2.8}$$

and $\otimes$ is a bilinear mapping. The linear hull of $U \times V$ being Q, $U \otimes V$ is the linear hull of $\otimes\,(U \times V)$. Finally, let $f \in L(U, V; W)$. The homomorphism $G: Q \to W$ be defined by

$$G\left(\sum_{i=1}^{r} a_i(u_i, v_i)\right) = \sum_{i=1}^{r} a_i f(u_i, v_i).$$

f being bilinear, we have $G(q) = 0$ for each q (2.7). Thus $G(N) = 0$. Let us define $g: U \otimes V \to W$ as follows: let $t \in Q/N$, $t' \in Q$ any element such that $\varkappa(t') = t$; set $g(t) = G(t')$. Evidently, this is a good definition. For $(u, v) \in U \times V$, $(g \circ \otimes)\,(u, v) = G\big((u, v)\big) = f\big((u, v)\big)$, i.e., the diagram (2.4) is commutative. QED.

It is usual to write $U \otimes V$ instead of $(U \otimes V, \otimes)$.

Theorem 2.3. *Let U, V, W be vector spaces. Then there are isomorphisms*

(2.9) $$\alpha: U \otimes V \to V \otimes U, \quad \beta: (U \otimes V) \otimes W \to U \otimes (V \otimes W), \quad \gamma: \mathbf{R} \otimes U \to U$$

such that

(2.10) $$\alpha(u \otimes v) = v \otimes u, \quad \beta[(u \otimes v) \otimes w] = u \otimes (v \otimes w), \quad \gamma(a \otimes u) = au; \quad u \in U, \quad v \in V, \quad w \in W, \quad a \in \mathbf{R}.$$

Proof. Consider the diagram

(2.11) $$\begin{array}{ccccc} & U \times V & \xrightarrow{\ \omega\ } & V \times U & \\ {\scriptstyle\otimes}\swarrow & & \searrow{\scriptstyle f_1}\quad {\scriptstyle\otimes}\swarrow & & \searrow{\scriptstyle f_2} \\ U \otimes V & \xrightarrow[\alpha_1]{} & V \otimes U & \xrightarrow[\alpha_2]{} & U \otimes V \end{array}$$

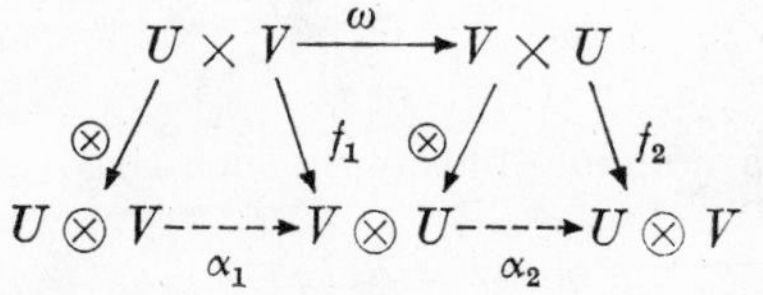

where $\omega(u, v) = (v, u)$, $f_1(u, v) = v \otimes u$ and $f_2(v, u) = u \otimes v$ for $u \in U$, $v \in V$. Obviously, $\otimes \circ \omega = f_1$. The mappings f_1, f_2 being bilinear, there are homomorphisms α_1, α_2 sucht that (2.11) is commutative. Because of $f_2 \circ \omega = \otimes: U \times V \to U \otimes V$ and the commutativity of (2.11), we get $\alpha_2 \circ \alpha_1 \circ \otimes = f_2 \circ \omega = \otimes$ and, consequently, $\alpha_2 \circ \alpha_1 = \mathrm{id}: U \otimes V \to U \otimes V$. Analogously, $\alpha_1 \circ \alpha_2 = \mathrm{id}$, and α_1, α_2 are isomorphisms, (2.10_1) being satisfied because of the definition of f_1. The isomorphisms β, γ are to be constructed analogously. QED.

Because of the last result, we may consider the tensor products of the form $V_1 \otimes \cdots \otimes V_r$, the position of the spaces V_i being irrelevant. We write $\otimes^k V = V \otimes \cdots \otimes V$ (k-times). The elements of the space $(\otimes^k V) \otimes (\otimes^l V^*)$ are called the *tensors* k-times *contravariant* and l-times *covariant*.

Theorem 2.4. *Let $f_1: U_1 \to V_1$, $f_2: U_2 \to V_2$ be homomorphisms. Then there is the homomorphism $f_1 \otimes f_2: U_1 \otimes U_2 \to V_1 \otimes V_2$ such that*

(2.12) $$(f_1 \otimes f_2)\,(u_1 \otimes u_2) = f_1(u_1) \otimes f_2(u_2); \quad u_1 \in U_1, \quad u_2 \in U_2.$$

Proof. Consider the diagram

$$
\begin{array}{ccc}
 & U_1 \times U_2 & \\
\otimes \swarrow & & \searrow f \\
U_1 \otimes U_2 & \xrightarrow[g]{} & V_1 \otimes V_2
\end{array} \tag{2.13}
$$

where $f(u_1, u_2) = f_1(u_1) \otimes f_2(u_2)$. The mapping f being bilinear, there is a homomorphism g such that (2.13) is commutative. Evidently, g is exactly the desired homomorphism $f_1 \otimes f_2$. Its unicity is the consequence of Theorem 2.1. QED.

Theorem 2.5. *Let U, V be vector spaces of finite dimensions, and $u_1, \ldots, u_m$ and $v_1, \ldots, v_n$ be the bases of U and V resp. Then $u_i \otimes v_\alpha$; $i = 1, \ldots, m$; $\alpha = 1, \ldots, n$; is a basis of $U \otimes V$.*

Proof. Consider the diagram

$$
\begin{array}{ccc}
 & U \times V & \\
\otimes \swarrow & & \searrow f_{k,\gamma} \\
U \otimes V & \xrightarrow[g_{k,\gamma}]{} & \mathbf{R}
\end{array} \tag{2.14}
$$

k, γ being fixed numbers and

$$f_{k,\gamma}\left(\sum_{i=1}^{m} A_i u_i, \sum_{\alpha=1}^{n} B_\alpha v_\alpha\right) = A_k B_\gamma.$$

The mapping $f_{k,\gamma}$ being bilinear, there is a homomorphism $g_{k,\gamma}$ such that

$$g_{k,\gamma}\left[\left(\sum_{i=1}^{m} A_i u_i\right) \otimes \left(\sum_{\alpha=1}^{n} B_\alpha v_\alpha\right)\right] = A_k B_\gamma.$$

Especially,

$$g_{k,\gamma}(u_i \otimes v_\alpha) = \begin{cases} 0 & \text{for} \quad k \neq i \quad \text{or for} \quad \gamma \neq \alpha \quad \text{resp.,} \\ 1 & \text{for} \quad k = i, \gamma = \alpha. \end{cases}$$

Suppose

$$\sum_{i=1}^{m} \sum_{\alpha=1}^{n} a_{i\alpha} u_i \otimes v_\alpha = 0, \quad a_{i\alpha} \in \mathbf{R}.$$

Applying $g_{k,\gamma}$, we get $a_{k\gamma} = 0$. QED.

3. Exterior forms

Let V be a vector space. Denote by Π_r the group of the permutations of the set $\{1, 2, \ldots, r\}$. Each element $\pi \in \Pi_r$ induces the isomorphism $\pi\colon \otimes^r V \to \otimes^r V$ characterized by the property

$$(3.1)\qquad \pi(v_1 \otimes \cdots \otimes v_r) = v_{\pi(1)} \otimes \cdots \otimes v_{\pi(r)}; \quad v_i \in V.$$

Definition 3.1. *A tensor* $T \in \otimes^r V$ *is called* antisymmetric *if*

$$(3.2)\qquad \pi(T) = (\operatorname{sign} \pi)\, T \quad \text{for each} \quad \pi \in \Pi_r.$$

A homomorphism $f\colon \otimes^r V \to W$ *is called* antisymmetric *if* $f \circ \pi = (\operatorname{sign} \pi) f$ *for each* $\pi \in \Pi_r$. *The homomorphism*

$$(3.3)\qquad A_r = \sum_{\pi \in \Pi_r} (\operatorname{sign} \pi)\, \pi\colon \otimes^r V \to \otimes^r V$$

is called the antisymmetrization.

Theorem 3.1. *Let* $T \in \otimes^r V$. *Then* $A_r(T)$ *is antisymmetric; T being antisymmetric, we have* $r!\, T = A_r(T)$.

Proof. Let $\pi_0 \in \Pi_r$. Then

$$\pi_0\big(A_r(T)\big) = \sum_{\pi \in \Pi_r} (\operatorname{sign} \pi)\, \pi_0\big(\pi(T)\big) = (\operatorname{sign} \pi_0)\, A_r(T).$$

For an antisymmetric T, we get

$$A_r(T) = \sum_{\pi \in \Pi_r} (\operatorname{sign} \pi)\, \pi(T) = \sum_{\pi \in \Pi_r} (\operatorname{sign} \pi)^2\, T = r!\, T.$$

QED.

Definition 3.2. *The* r-th exterior product *of the vector space V is the space*

$$(3.4)\qquad \wedge^r V = \otimes^r V / \operatorname{Ker} A_r.$$

The image of $v_1 \otimes \cdots \otimes v_r \in \otimes^r V$ *in the natural homomorphism* $\otimes^r V \to \wedge^r V$ *is denoted by* $v_1 \wedge \cdots \wedge v_r$. *For* $r = 0$, *define* $\wedge^0 V = \mathbf{R}$.

The easy proofs of the following theorems are left to the reader.

Theorem 3.2. *Let* $f\colon \otimes^r V \to W$ *be an antisymmetric homomorphism. Then there is the homomorphism g such that the diagram*

$$(3.5)\qquad \begin{array}{ccc} & \otimes^r V & \\ {\scriptstyle\mu}\swarrow & & \searrow{\scriptstyle f} \\ \wedge^r V & \xrightarrow[g]{} & W \end{array}$$

is commutative. If $A(\otimes^r V; W) \subset L(\otimes^r V; W)$ *denotes the subspace of antisymmetric homomorphisms, the mapping* $A(\otimes^r V; W) \to L(\wedge^r V; W)$, $f \mapsto g$, *is an isomorphism.*

Theorem 3.3. *There is a unique monomorphism* $\varkappa \equiv \varkappa_r\colon \wedge^r V \to \otimes^r V$ *such that*

$$(3.6)\qquad \varkappa(v_1 \wedge \cdots \wedge v_r) = A_r(v_1 \otimes \cdots \otimes v_r), \quad v_i \in V.$$

$\varkappa$ *is an isomorphism between* $\wedge^r V$ *and the vector subspace of all antisymmetric tensors of* $\otimes^r V$.

Definition 3.3. *The* exterior multiplication *is the bilinear mapping*

$$\wedge \colon \wedge^r V \times \wedge^s V \to \wedge^{r+s} V \tag{3.7}$$

given by the condition

$$T_1 \wedge T_2 = \frac{1}{r!\,s!}\, \mu_{r+s}\big(\varkappa_r(T_1) \otimes \varkappa_s(T_2)\big); \quad T_1 \in \wedge^r V, \quad T_2 \in \wedge^s V; \tag{3.8}$$

$\varkappa_r$, $\varkappa_s$ being the monomorphisms of Theorem 3.3 *and $\mu_{r+s}\colon \otimes^{r+s} V \to \wedge^{r+s} V$ the natural mapping.*

Theorem 3.4. *We have*

$$\begin{aligned} T_1 \wedge T_2 &= (-1)^{rs}\, T_2 \wedge T_1, \quad (T_1 \wedge T_2) \wedge T_3 = T_1 \wedge (T_2 \wedge T_3),\\ T_1 \wedge (T_2 + T_2') &= T_1 \wedge T_2 + T_1 \wedge T_2' \end{aligned} \tag{3.9}$$

for $T_1 \in \wedge^r V$; $T_2, T_2' \in \wedge^s V$; $T_3 \in \wedge^t V$.

Theorem 3.5. *To each homomorphism $f\colon V \to W$, there is the homomorphism $\wedge^r f\colon \wedge^r V \to \wedge^r W$ characterized by the property*

$$(\wedge^r f)\,(v_1 \wedge \cdots \wedge v_r) = f(v_1) \wedge \cdots \wedge f(v_r) \text{ for all } v_i \in V. \tag{3.10}$$

Theorem 3.6. *Let $v_1, \ldots, v_n$ be a basis of V. Then the elements*

$$v_{i_1} \wedge \cdots \wedge v_{i_r}; \quad 1 \leqq i_1 < i_2 < \cdots < i_r \leqq n; \tag{3.11}$$

form a basis of $\wedge^r V$.

Suppose $\dim V = n$; then, see (3.9_1) and Theorem 3.6, $\wedge^m V = 0$ for $m > n$ and $\dim \wedge^m V = \binom{n}{m}$ for $0 \leqq m \leqq n$. Each element $\xi \in \wedge^r V$ may be written as

$$\xi = \sum_{1 \leqq i_1 < \cdots < i_r \leqq n} a_{i_1 \ldots i_r} v_{i_1} \wedge \cdots \wedge v_{i_r}. \tag{3.12}$$

It is often much more convenient to write the same element as

$$\xi = \frac{1}{r!} \sum_{i_1=1}^{n} \cdots \sum_{i_r=1}^{n} b_{i_1 \ldots i_r} v_{i_1} \wedge \cdots \wedge v_{i_r}, \tag{3.13}$$

the coefficients $b_{i_1 \ldots i_r}$ satisfying the condition

$$b_{i_{\pi(1)} \cdots i_{\pi(r)}} = (\operatorname{sign} \pi)\, b_{i_1 \ldots i_r} \quad \text{for each} \quad \pi \in \Pi_r. \tag{3.14}$$

From this,

$$b_{i_1 \ldots i_r} = \begin{cases} 0 & \text{if at least two numbers } i_1, \ldots, i_r \text{ are equal,}\\ (\operatorname{sign} \pi)\, a_{i_{\pi(1)} \cdots i_{\pi(r)}}, & \pi \in \Pi_r \text{ satisfying } i_{\pi(1)} < \cdots < i_{\pi(r)}. \end{cases}$$

Theorem 3.7. *Let U, V be finite-dimensional vector spaces. Then there exists an isomorphism*

$$\sigma\colon V \otimes U^* \to L(U; V) \tag{3.15}$$

such that

(3.16) $[\sigma(v \otimes u^*)](u) = u^*(u) \cdot v$ *for each* $u \in U, v \in V, u^* \in U^*$.

Proof. For $v \in V$, $u^* \in U^*$, define the homomorphism $\tau_{v,u^*}: U \to V$ by the relation $\tau_{v,u^*}(u) = u^*(u) \cdot v$. Then the mapping $V \times U^* \to L(U; V)$, $(v, u^*) \mapsto \tau_{v,u^*}$, is bilinear. Thus there is a homomorphism $\sigma: V \otimes U^* \to L(U, V)$ such that $\sigma(v \otimes u^*) = \tau_{v,u^*}$. Let $v_1, \ldots, v_n$ be a basis of V, $u_1, \ldots, u_m$ be a basis of U. Because of $\dim (V \otimes U^*) = \dim L(U; V)$, it is sufficient to prove that the elements τ_{v^i,u^α} are linearly independent; here, v^i and u^α are the dual bases resp. Suppose

$$\sum_{i=1}^{n} \sum_{\alpha=1}^{m} a_{i\alpha} \tau_{v_i,u^\alpha} = 0.$$

Then $\tau_{v_i,u^\alpha}(u_\beta) = u^\alpha(u_\beta)\, v_i = \delta_\beta{}^\alpha v_i$, and we get

$$\sum_{i=1}^{n} \sum_{\alpha=1}^{m} a_{i\alpha} \delta_\beta{}^\alpha v_i = \sum_{i=1}^{n} a_{i\beta} v_i = 0.$$

Thus $a_{i\beta} = 0$ for $i = 1, \ldots, n$; $\beta = 1, \ldots, m$. QED.

Theorem 3.8. *Let U, V be finite-dimensional vector spaces. Then there is an isomorphism*

(3.17) $\tau: U^* \otimes V^* \to (U \otimes V)^*$

with the following property: let $u^* \in U^*$, $v^* \in V^*$, then $\tau(u^* \otimes v^*) \in (U \times V)^*$ equals to $u^* \otimes v^*: U \otimes V \to \mathbf{R} \otimes \mathbf{R} \cong \mathbf{R}$, the identification $\mathbf{R} \otimes \mathbf{R} \cong \mathbf{R}$ being constructed by means of the isomorphism $\gamma: \mathbf{R} \otimes \mathbf{R} \to \mathbf{R}$ of Theorem 2.3.

Proof. Let $\tau_1: U^* \times V^* \to (U \otimes V)^*$ satisfy the analogous property: $\tau_1(u^*, v^*): U \otimes V \to \mathbf{R}$ be equal to the homomorphism $u^* \otimes v^*: U \otimes V \to \mathbf{R} \otimes \mathbf{R} \cong \mathbf{R}$. The mapping τ_1 being bilinear, the existence of τ follows. QED.

Theorem 3.8 may be generalized as follows.

Theorem 3.9. *Let $V_1, \ldots, V_r$ be finite-dimensional vector spaces. Then there is an isomorphism*

(3.18) $\tau: V_1^* \otimes \cdots \otimes V_r^* \to (V_1 \otimes \cdots \otimes V_r)^*$

with the following property: let $v_i^* \in V_i^*$; then $\tau(v_1^* \otimes \cdots \otimes v_r^*) \in (V_1 \otimes \cdots \otimes V_r)^*$ is exactly the homomorphism $v_1^* \otimes \cdots \otimes v_r^*: V_1 \otimes \cdots \otimes V_r \to \mathbf{R} \otimes \cdots \otimes \mathbf{R} \cong \mathbf{R}$.

Theorem 3.10. *Let V be a finite-dimensional vector space. Then there is an isomorphism*

(3.19) $\mu: \wedge^r V^* \to (\wedge^r V)^*$

such that

(3.20) $[\mu(v^1 \wedge \cdots \wedge v^r)](w_1 \wedge \cdots \wedge w_r) = \det \|v^i(w_j)\|$ *for* $v^i \in V^*$, $w_j \in V$.

Proof. Let $\tau: \otimes^r V^* \to (\otimes^r V)^*$ be the isomorphism (3.18); further, let $A^r \subset \otimes^r V^*$ be the subspace of antisymmetric tensors and $A(\otimes^r V; \mathbf{R}) \subset L(\otimes^r V; \mathbf{R}) = (\otimes^r V)^*$

the subspace of antisymmetric homomorphisms. First of all, let us prove

$$(3.21)\qquad \tau\big(\pi(T^*)\big) = \tau(T^*) \circ \pi^{-1} \quad \text{for} \quad T^* \in \otimes^r V^*, \quad \pi \in \Pi_r .$$

Suppose $T^* = v^1 \otimes \cdots \otimes v^r$, $v^i \in V^*$; $w_1, \ldots, w_r \in V$ be arbitrary. Then

$$\big[\tau\big(\pi(v^1 \otimes \cdots \otimes v^r)\big)\big]\,(w_1 \otimes \cdots \otimes w_r) = [\tau(v^{\pi(1)} \otimes \cdots \otimes v^{\pi(r)})]\,(w_1 \otimes \cdots \otimes w_r)$$
$$= v^{\pi(1)}(w_1) \cdots v^{\pi(r)}(w_r).$$

Further,

$$[\tau(v^1 \otimes \cdots \otimes v^r) \circ \pi^{-1}]\,(w_1 \otimes \cdots \otimes w_r)$$
$$= [\tau(v^1 \otimes \cdots \otimes v^r)]\,(w_{\pi^{-1}(1)} \otimes \cdots \otimes w_{\pi^{-1}(r)})$$
$$= v^1(w_{\pi^{-1}(1)}) \cdots v^r(w_{\pi^{-1}(r)}) = v^{\pi(1)}(w_1) \cdots v^{\pi(r)}(w_r).$$

Thus (3.21) holds true for any tensor T^* of the form $v^1 \otimes \cdots \otimes v^r$, and, consequently, for any tensor T^*. From this, T^* is antisymmetric if and only if the corresponding homomorphism $\tau(T^*)\colon \otimes^r V \to \mathbf{R}$ is antisymmetric. Restricting the isomorphism τ, we get the isomorphism $\tau\colon A^r \to A(\otimes^r V; \mathbf{R})$.

Consider the sequence

$$(3.22)\qquad \wedge^r V^* \xrightarrow{\varkappa} A^r \xrightarrow{\tau} A(\otimes^r V; \mathbf{R}) \xrightarrow{\lambda} L(\wedge^r V; \mathbf{R}) = (\wedge^r V)^*,$$

τ being just constructed, $\varkappa$ and λ being the isomorphisms of Theorems 3.3 and 3.2 resp. Set $\mu = \lambda \circ \tau \circ \varkappa$. Let $t = v^1 \wedge \cdots \wedge v^r \in \wedge^r V^*$. Then

$$\varkappa(t) = A_r(v^1 \otimes \cdots \otimes v^r) = \sum_{\pi \in \Pi_r} (\operatorname{sign} \pi)\,(v^{\pi(1)} \otimes \cdots \otimes v^{\pi(r)})$$

and

$$\big[\tau\big(\varkappa(t)\big)\big]\,(w_1 \otimes \cdots \otimes w_r) = \sum_{\pi \in \Pi_r} (\operatorname{sign} \pi)\, v^{\pi(1)}(w_1) \cdots v^{\pi(r)}(w_r) = \det \|v^i(w_j)\|.$$

Applying λ, we get exactly (3.20). QED.

Definition 3.4. *Let V be a vector space. The space $\wedge^r V^*$ is called the* space of exterior r-forms on V. *We set $\wedge^0 V^* = \mathbf{R}$.*

The sequence (3.22) shows the spaces isomorphic to $\wedge^r V^*$. The most interesting and usefull is the space $A(\otimes^r V; \mathbf{R})$. Consider the space $L(V, \ldots, V; \mathbf{R}) = L(\times^r V; \mathbf{R})$. A function $f \in L(\times^r V; \mathbf{R})$ is called *antisymmetric if*

$$(3.23)\qquad f(v_1, \ldots, v_i, \ldots, v_j, \ldots, v_r) = -f(v_1, \ldots, v_j, \ldots, v_i, \ldots, v_r)$$
$$\text{for each } 1 \leqq i < j \leqq r.$$

Let $A(\times^r V; \mathbf{R}) \subset L(\times^r V; \mathbf{R})$ be the subspace of all antisymmetric multilinear functions. The following result is immediate.

Theorem 3.11. *To each function $f \in A(\times^r V; \mathbf{R})$ there is exactly one function $g \in A(\otimes^r V; \mathbf{R})$ such that the diagram*

$$(3.24)\qquad \begin{array}{ccc} & \times^r V & \\ \otimes \swarrow & & \searrow f \\ \otimes^r V & \xrightarrow[g]{} & \mathbf{R} \end{array}$$

is commutative. The mapping

$$(3.25)\qquad \nu\colon A(\times^r V;\mathbf{R}) \to A(\otimes^r V;\mathbf{R}),\quad f \mapsto g,$$

is an isomorphism.

Thus we get the isomorphism

$$(3.26)\qquad \iota = \nu^{-1}\circ\tau\circ\varkappa\colon \wedge^r V^* \to A(\times^r V;\mathbf{R}).$$

Let us exhibit it quite explicitely. Let $v^1, \ldots, v^r \in V^*$ and $\omega = v^1 \wedge \cdots \wedge v^r \in \wedge^r V^*$; further, let $w_1, \ldots, w_r \in V$; we have to calculate $[\iota(\omega)]\,(w_1, \ldots, w_r)$. But this is easy: from

$$\varkappa(\omega) = A_r(\omega) = \sum_{\pi\in\Pi_r} (\operatorname{sign}\pi)\, v^{\pi(1)} \otimes \cdots \otimes v^{\pi(r)}$$

and

$$[(\tau\circ\varkappa)\,(\omega)]\,(w_1 \otimes \cdots \otimes w_r) = \det \|v^i(w_j)\|,$$

we get

$$(3.27)\qquad [\iota(v^1 \wedge \cdots \wedge v^r)]\,(w_1, \ldots, w_r) = \det \|v^i(w_j)\|.$$

Considering a general r-form

$$(3.28)\qquad \omega = \frac{1}{r!}\sum_{i_1=1}^{n} \cdots \sum_{i_r=1}^{n} b_{i_1\ldots i_r} v^{i_1} \wedge \cdots \wedge v^{i_r},$$

$v^1, \ldots, v^n$ being the basis of V^*, we have

$$(3.29)\qquad [\iota(\omega)]\,(w_1, \ldots, w_r) = \frac{1}{r!}\sum_{i_1=1}^{n} \cdots \sum_{i_r=1}^{n} b_{i_1\ldots i_r} \begin{vmatrix} v^{i_1}(w_1), & \ldots, & v^{i_r}(w_1) \\ \cdot & \cdots & \cdot \\ v^{i_1}(w_r), & \ldots, & v^{i_r}(w_r) \end{vmatrix}.$$

Theorem 3.12. *Let $\omega_1 \in \wedge^r V^*$, $\omega_2 \in \wedge^s V^*$; $w_1, \ldots, w_{r+s} \in V$. Then*

$$(3.30)\qquad [\iota(\omega_1 \wedge \omega_2)]\,(w_1, \ldots, w_{r+s})$$
$$= \frac{1}{r!s!}\sum_{\pi\in\Pi_{r+s}} (\operatorname{sign}\pi)\,[\iota(\omega_1)]\,(w_{\pi(1)}, \ldots, w_{\pi(r)}) \cdot [\iota(\omega_2)]\,(w_{\pi(r+1)}, \ldots, w_{\pi(r+s)}).$$

Proof. Let $v_1, \ldots, v_n$ be a basis of V and $v^1, \ldots, v^n$ the dual basis. First of all, let $\omega_1 = v^1 \wedge \cdots \wedge v^r$, $\omega_2 = v^{r+1} \wedge \cdots \wedge v^{r+s}$, $(w_1, \ldots, w_{r+s}) = (v_1, \ldots, v_{r+s})$. Then $[\iota(\omega_1 \wedge \omega_2)]\,(w_1, \ldots, w_{r+s}) = 1$. Further, $[\iota(\omega_1)]\,(w_{\pi(1)}, \ldots, w_{\pi(r)}) \neq 0$, $\pi \in \Pi_{r+s}$, if and only if $\pi(1), \ldots, \pi(r)$ is a permutation of $1, \ldots, r$; analogously, for $[\iota(\omega_2)]\,(w_{\pi(r+1)}, \ldots, w_{\pi(r+s)})$. The right-hand side of (3.30) is thus $\neq 0$ if and only if π is a permutation permuting the initial r numbers. Thus the right-hand side is equal to

$$\frac{1}{r!s!}\sum_{\pi'\in\Pi_r}\sum_{\pi''\in\Pi_s} (\operatorname{sign}\pi')\,(\operatorname{sign}\pi'')\,[\iota(\omega_1)]\,(v_{\pi'(1)}, \ldots, v_{\pi'(r)}).$$

$$[\iota(\omega_2)]\,(v_{\pi''(r+1)}, \ldots, v_{\pi''(r+s)}) = \frac{1}{r!s!}\sum_{\pi'\in\Pi_r}\sum_{\pi''\in\Pi_s} (\operatorname{sign}\pi')^2\,(\operatorname{sign}\pi'')^2 = 1.$$

$(w_1, \ldots, w_{r+s})$ being another sequence of the elements of the basis, both sides of (3.30) are equal to zero. Finally, ω_1 and ω_2 being exterior products of the elements of the basis, one of these elements being contained in both ω_1 and ω_2, both sides of (3.30) are equal to zero as well. Thus (3.30) follows from the linearity. QED.

In what follows, we are going to identify the spaces $\wedge^r V^*$ and $A(\times^r V; \mathbf{R})$ by means of ι.

Finally, let us prove a very usefull auxiliary result (the so-called Cartan's lemma):

Theorem 3.13. *Let* $v_1, \ldots, v_r \in V$ *be independent vectors,* $u_1, \ldots, u_r \in V$, *and let*

$$(3.31) \qquad v_1 \wedge u_1 + \cdots + v_r \wedge u_r = 0.$$

Then there are $a_{ij} = a_{ji} \in \mathbf{R}$; $i, j = 1, \ldots, r$; *such that*

$$(3.32) \qquad u_i = \sum_{j=1}^{r} a_{ij} v_j.$$

Proof. Let us complete $v_1, \ldots, v_r$ to a basis $v_1, \ldots, v_r, v_{r+1}, \ldots, v_{r+s}$ of V. Then

$$u_i = \sum_{j=1}^{r} a_{ij} v_j + \sum_{\alpha=1}^{s} b_{i\alpha} v_{r+\alpha}.$$

Substituting into (3.31), we get

$$\sum_{i=1}^{r} \sum_{j=1}^{r} a_{ij} v_i \wedge v_j + \sum_{i=1}^{r} \sum_{\alpha=1}^{s} b_{i\alpha} v_i \wedge v_{r+\alpha} = 0.$$

From Theorem 3.6, $a_{ij} = a_{ji}$, $b_{i\alpha} = 0$. QED.

II. Differentiable manifolds

1. Differentiable manifolds and mappings

First of all, let us recall some elementary topological definitions.

The *topological space* is a set X with a given collection $\mathcal{T}$ of (so-called *open*) subsets with the following properties: (i) $G_1, G_2 \in \mathcal{T}$ implies $G_1 \cap G_2 \in \mathcal{T}$; (ii) I being any system of indices, $G_\alpha \in \mathcal{T}$, $\alpha \in I$ implies $\bigcup_{\alpha \in I} G_\alpha \in \mathcal{T}$; (iii) $\emptyset \in \mathcal{T}$, $X \in \mathcal{T}$. For $A \subset X$, its *interior* is defined as $\operatorname{Int} A = \{\bigcup B; B \subset A, B \in \mathcal{T}\}$. The *neighbourhood* of $x \in X$ is any subset $A \subset X$ such that $x \in \operatorname{Int} A$ (in what follows, we are going to restrict ourselves to open neighbourhoods). The subset $A \subset X$ is called *closed,* if $X - A$ is open. The *closure* $\overline{B}$ of B is defined as the intersection of all closed sets in X containing B. The topological space X is called *Hausdorff* if, for each couple of points $x, y \in X$, $x \neq y$, there are its neighbourhoods A_x, A_y such that $A_x \cap A_y = \emptyset$. A system $\mathcal{B}$ of open sets of X is called the *basis* of X if each open set is a union of elements of $\mathcal{B}$. A system $\mathcal{B}$ of subsets of a set X is a basis of a topology $\mathcal{T}$ on X if X is the union of elements of $\mathcal{B}$ and the intersection of a finite number of elements of $\mathcal{B}$ is a union of elements of $\mathcal{B}$; the open subsets in $\mathcal{T}$ are to be defined as all possible unions of sets from $\mathcal{B}$. Let $A \subset X$; the topology $\mathcal{T}$ on X induces a topology $\mathcal{T}_A = \{G \cap A; G \in \mathcal{T}\}$ on A. A *domain* will be a connected open set.

Let X_α, $\alpha \in I$, be a system of topological spaces with topologies $\mathcal{T}_\alpha$. Let X be the system of all functions $x\colon I = \bigcup_{\alpha \in I} X_\alpha$ such that $x(\alpha) \in X_\alpha$; X is the so-called cartesian product of the system X_α. For each $\alpha \in I$, define the mapping $p\colon X \to X_\alpha$ by the formula $p_\alpha(x) = x(\alpha)$. Let S be the system of all sets of the form $p_\alpha^{-1}(U_\alpha)$, $U_\alpha \in \mathcal{T}_\alpha$. The system of all possible intersections of finite numbers of elements from S is then a basis of a topology $\mathcal{T}$ on X; $\mathcal{T}$ is the so called *product topology.*

Let X, Y be topological spaces. The mapping $f\colon X \to Y$ is called *continuous* if $f^{-1}(G) \in \mathcal{T}_X$ for each $G \in \mathcal{T}_Y$; it is called a *homeomorphism* if it is one-to-one and both f and f^{-1} are continuous.

The *metric* on a set X is a function $d\colon X \times X \to \mathbf{R}$ with the following properties: (i) $d(x, y) \geqq 0$ and $d(x, y) = 0$ if and only if $x = y$; (ii) $d(x, y) = d(y, x)$; (iii) $d(x, y) + d(y, z) \geqq d(x, z)$. A metric d on X induces a topology $\mathcal{T}$ on X: $G \in \mathcal{T}$ if and only if to each point $x \in G$ there is a number $\varepsilon_x > 0$ such that $\{y \in X; d(x, y) < \varepsilon_x\} \subset G$. The *standard topology* of $\mathbf{R}^n$ is induced by the metric $d(x, y) = \max(|x^1 - y^1|, \ldots, |x^n - y^n|)$; $x = (x^1, \ldots, x^n)$, $y = (y^1, \ldots, y^n)$.

Definition 1.1. *Let $V \subset \mathbf{R}^n$ be an open set (in the standard topology). The function $\varphi\colon V \to \mathbf{R}$ is of* class C^k $(k = 0, 1, \ldots)$ *if there exist its continuous derivatives up to the order k inclusive; φ is of* class C^∞ *if it is of class C^k for any k; φ is of* class C^ω (or holomorphic) *if φ may be expressed, in a suitable neighbourhood of each point $v \in V$,* *as a power series centered at* v. *Let* $\mathrm{pr}^i\colon \mathbf{R}^n \to \mathbf{R}$ *be defined by* $\mathrm{pr}^i(x^1, \ldots, x^n) = x^i$; *the mapping* $\psi\colon W \to \mathbf{R}^n$ (W *being an open subset of* $\mathbf{R}^n$) *is called of* class C^k ($k = 0, 1, \ldots, \infty, \omega$) *if all mappings* $\mathrm{pr}^i \circ \psi\colon W \to \mathbf{R}$ *are of class* C^k.

Definition 1.2. *Let* X *be a topological space. A* map *of* X *is a couple* (U, μ) *with* U *an open subset of* X *and* $\mu\colon U \to \mathbf{R}^n$ *a homeomorphism between* U *and* $\mu(U)$; *the* dimension *of* (U, μ) *is the number* n, *the* coordinates *of* (U, μ) *are the functions* $\mathrm{pr}^i \circ \mu\colon U \to \mathbf{R}$. *The neighbourhood* U *of* $x \in X$ *is said to be a* coordinate neighbourhood *of* x *if there is a map* $\mu\colon U \to \mathbf{R}^n$; *its coordinates are called the* local coordinates *around* x. *Two maps* $\mu\colon U \to \mathbf{R}^n$, $\tau\colon V \to \mathbf{R}^m$ *of* X *are called* C^k-related *if* $n = m$ *and* $U \cap V = \emptyset$ *or* $U \cap V \neq \emptyset$ *and the mappings* $\mu \circ \tau^{-1}\colon \tau(U \cap V) \to \mu(U \cap V)$, $\tau \circ \mu^{-1}\colon \mu(U \cap V) \to \tau(U \cap V)$ *are of class* C^k. *An* atlas $\mathscr{A}$ of X *is a system of maps* $\{\mu_\alpha\colon U_\alpha \to \mathbf{R}^{n_\alpha};\ \alpha \in I\}$ *such that* $\bigcup_{\alpha \in I} U_\alpha = X$. *The atlas* $\mathscr{A}$ *is called of* class C^k *if any its maps* μ_α, μ_β *are* C^k-*related. The map* $\mu\colon U \to \mathbf{R}^m$ *is said to be* C^k-related *with* $\mathscr{A}$ *if it is* C^k-*related with all its maps. The atlas* $\mathscr{A}$ *of class* C^k *is called* maximal *if it contains each map* C^k-*related with it.*

Definition 1.3. *The* differentiable manifold M of class C^k (*or a* C^k-manifold) *is a connected Hausdorff topological space with a countable basis together with a maximal atlas of class* C^k. *The* dimension *of* M *is defined as the dimension of any map of its atlas.*

To give a differentiable structure on X, it is sufficient to give just an atlas. This follows from the following trivial assertion: *Be given an atlas* $\mathscr{A}'$ *of* X *of class* C^k; *then there is a unique maximal atlas* $\mathscr{A}$ *of* X *of class* C^k *such that* $\mathscr{A}' \subset \mathscr{A}$.

Theorem 1.1. *Let* M, N *be* C^k-*manifolds of dimensions* m *and* n *resp.; further, let* $\{\mu_\alpha\colon U_\alpha \to \mathbf{R}^n;\ \alpha \in I\}$, $\{\nu_\beta\colon V_\beta \to \mathbf{R}^m;\ \beta \in J\}$ *be the corresponding atlases. Then*

$$\{(\mu_\alpha, \nu_\beta)\colon U_\alpha \times V_\beta \to \mathbf{R}^n \times \mathbf{R}^m = \mathbf{R}^{n+m};\ \alpha \in I, \beta \in J\};$$

$$(\mu_\alpha, \nu_\beta)\,(x, y) = \big(\mu_\alpha(x), \nu_\beta(y)\big);$$

is a C^k-*atlas on* $M \times N$.

The proof of this theorem is trivial. Because of it, $M \times N$ becomes a C^k-manifold with $\dim\,(M \times N) = \dim M \cdot \dim N$.

Definition 1.4. *Let* M *be a* C^k-*manifold and* $f\colon M \to \mathbf{R}^p$ *a mapping.* f *is called of* class C^l ($l \leqq k$) *at the point* $m \in M$ *if for a map (and then for any map)* $\mu_\alpha\colon U_\alpha \to \mathbf{R}^n$, $m \in U_\alpha$, *the mapping* $f \circ \mu_\alpha^{-1}\colon \mu_\alpha(U_\alpha) \to \mathbf{R}^p$ *is of class* C^l *at* $\mu_\alpha(x)$; f *is of* class C^l *on* $G \subset M$ *if it is of class* C^l *at each point* $m \in G$. *Let* M, N *be* C^k-*manifolds and* $f\colon M \to N$ *a mapping.* f *is of* class C^l ($l \leqq k$) *at* $m \in M$ *if it is continuous and we have:* $\nu_\beta\colon V_\beta \to \mathbf{R}^n$ *being a map of* N *at* $f(m) \in N$, *the mapping* $\nu_\beta \circ f\colon f^{-1}(V_\beta) \to \mathbf{R}^m$ *is of class* C^l *at* m. *The* C^k-diffeomorphism *of* M *onto* N *is a one-to-one mapping* $f\colon M \to N$ *of class* C^k *such that* $f^{-1}\colon N \to M$ *is of class* C^k *as well.*

In what follows, *all manifolds and mappings are considered to be of class* C^∞.

2. Tangent vectors

Definition 2.1. *Let M and N be manifolds and $m \in M$. The mappings $f_1: U_1 \to N$, $f_2: U_2 \to N$, U_1 and U_2 being neighbourhoods of m, are called* equivalent *at m if there is a neighbourhood U of m such that $U \subset U_1 \cap U_2$ and $f_1|_U = f_2|_U$. Each class of this equivalence is called the* germ. *The* germ of f, *defined in a neighbourhood of $m \in M$, is the equivalence class of mappings containing f.*

For a manifold M and $m \in M$, denote by $\mathscr{F}^\infty(m) \equiv \mathscr{F}^\infty(m, M)$ the set of germs of all C^∞-functions $f: U \to \mathbf{R}$ defined in the neighbourhoods of m; let us write $f \in \mathscr{F}^\infty(m)$ instead of germ $f \in \mathscr{F}^\infty(m)$. Let $\alpha, \beta \in \mathscr{F}^\infty(m, M)$; $r_1, r_2 \in \mathbf{R}$. Let $f: U_1 \to \mathbf{R}$, $g: U_2 \to \mathbf{R}$ be functions such that $\alpha =$ germ f, $\beta =$ germ g; let $U = U_1 \cap U_2$. *The set* $\mathscr{F}^\infty(m, M)$ *is* made into *a ring* by means of the following definitions:

$$r_1\alpha + r_2\beta = \text{germ}\,(r_1 f|_U + r_2 g|_U), \quad \alpha\beta = \text{germ}\,(f|_U \cdot g|_U).$$

For $\alpha \in \mathscr{F}^\infty(x)$, define $\alpha(m) = f(m)$, f being any function such that germ $f = \alpha$.

Definition 2.2. *The* tangent vector *at $m \in M$ is each mapping $t: \mathscr{F}^\infty(m) \to \mathbf{R}$ satisfying*

$$\text{(i) } t(r_1\alpha + r_2\beta) = r_1 t(\alpha) + r_2 t(\beta) \quad \textit{for} \quad r_1, r_2 \in \mathbf{R};\ \alpha, \beta \in \mathscr{F}^\infty(m);$$

$$\text{(ii) } t(\alpha\beta) = \beta(m)\, t(\alpha) + \alpha(m)\, t(\beta) \quad \textit{for} \quad \alpha, \beta \in \mathscr{F}^\infty(m).$$

The set of all tangent vectors at $m \in M$ is called the tangent space $T_m(M)$. *If f is a function defined on a neighbourhood of m and $t \in T_m(M)$, we define* $t(f) = t\,(\text{germ}\, f)$.

$T_m(M)$ is a vector space if we define $(r_1t_1 + r_2t_2)\,(\alpha) = r_1t_1(\alpha) + r_2t_2(\alpha)$.

Theorem 2.1. *Let M be a manifold, $\mu: U \to \mathbf{R}^n$ a map of M, $m \in U$. Let $(x^1, \ldots, x^n)$ be coordinates in $\mathbf{R}^n$. The mappings*

$$\text{(2.1)} \qquad \left.\frac{\partial}{\partial x^i}\right|_m : \mathscr{F}^\infty(m, M) \to \mathbf{R}; \quad i = 1, \ldots, n;$$

be defined by

$$\text{(2.2)} \qquad \left.\frac{\partial}{\partial x^i}\right|_m (\alpha) = \left.\frac{\partial(f \circ \mu^{-1})}{\partial x^i}\right|_{\mu(m)} \qquad \textit{for} \quad \text{germ}\, f = \alpha,\ f: U \to \mathbf{R}.$$

The mappings (2.1) *are then linearly independent tangent vectors of M at m.*

Proof. Obviously, (2.2) does not depend on the choice of $f \in$ germ f. It is also easy to see that (2.1) are tangent vectors. Now, let

$$t = \sum_{i=1}^{n} a^i \cdot \left.\frac{\partial}{\partial x^i}\right|_m; \qquad a^1, \ldots, a^n \in \mathbf{R}.$$

Consider the function $f_1: U \to \mathbf{R}$ defined by $f_j(y) = (\mathrm{pr}^j \circ \mu)\,(y)$. Then it yields $(f_j \circ \mu^{-1})\,(x^1, \ldots, x^n) = x^j$, and $\partial/\partial x^i|_m(f_j) = \delta_i{}^j$. Thus $t(f_j) = a^j$. From $t = 0$, we get $a^j = 0$, $j = 1, \ldots, n$, this proving the linear independency of vectors (2.1). QED.

Usually, we are going to write $\partial f(m)/\partial x^i$ instead of $\partial/\partial x^i|_m$ (germ f).

Theorem 2.2. *Let M be a manifold, $\mu\colon U \to \mathbf{R}^n$ its map, $m \in U$, $t \in T_m(M)$. Then there are (uniquely determined) numbers $a^1, \ldots, a^n \in \mathbf{R}$ such that*

$$(2.3) \qquad t = \sum_{i=1}^{n} a^i \cdot \left.\frac{\partial}{\partial x^i}\right|_y .$$

Thus $\dim T_m(M) = \dim M$.

Proof. Without loss of generality, let us suppose $\mu(m) = (0, \ldots, 0) \in \mathbf{R}^n$.

(1) Be given $\alpha \in \mathscr{F}^\infty(m)$. Choose $f\colon U \to \mathbf{R}$ such that $\alpha = \operatorname{germ} f$. Suppose that the function $\bar{f}\colon \mu(U) \to \mathbf{R}$, $\bar{f} = f \circ \mu^{-1}$, may be written as

$$(2.4) \qquad \bar{f}(x^1, \ldots, x^n) = \bar{f}(0, \ldots, 0) + \sum_{i=1}^{n} x^i \bar{f}_i(x^1, \ldots, x^n),$$

$\bar{f}_i$ being certain functions defined in the neighbourhood of $\mu(m)$. From (2.2),

$$\frac{\partial f(m)}{\partial x^j} = \bar{f}_j(0, \ldots, 0).$$

Obviously,

$$f(x) = (\bar{f} \circ \mu)(x) = \bar{f}\big(\mu(x)\big) = \bar{f}(0, \ldots, 0) + \sum_{i=1}^{n} (\mathrm{pr}^i \circ \mu)(x) \cdot \bar{f}_i\big(\mu(x)\big)$$

and

$$t(f) = \sum_{i=1}^{n} [t(\mathrm{pr}^i \circ \mu) \cdot \bar{f}_i(0, \ldots, 0) + (\mathrm{pr}^i \circ \mu)(m) \cdot t(\bar{f}_i \circ \mu)] = \sum_{i=1}^{n} t(\mathrm{pr}^i \circ \mu) \frac{\partial f(m)}{\partial x^i}$$

because of $(\mathrm{pr}^i \circ \mu)(y) = 0$ and $t\big(\bar{f}(0, \ldots, 0)\big) = 0$; from this we get $a^i = t(\mathrm{pr}^i \circ \mu)$.

(2) It is sufficient to prove that, to a given function $\bar{f}$ defined on a neighbourhood V of $(0, \ldots, 0) \in \mathbf{R}^n$, there exist (on a perhaps smaller neighbourhood $V_1 \subset V$) functions $\bar{f}_i$ with the property (2.4) on V_1. Consider a neighbourhood V_1 of $(0, \ldots, 0)$ with the property that $(x^1, \ldots, x^n) \in V_1$ implies $(tx^1, \ldots, tx^n) \in V_1$ for each t, $0 \leqq t \leqq 1$. Take a fixed point $(p^1, \ldots, p^n) \in V_1$. Then $\bar{f}(tp^1, \ldots, tp^n)$ is defined for $0 \leqq t \leqq 1$, and we have

$$\frac{\mathrm{d}\bar{f}(tp^1, \ldots, tp^n)}{\mathrm{d}t} = \sum_{i=1}^{n} \frac{\partial \bar{f}(tp^1, \ldots, tp^n)}{\partial x^i} p^i .$$

From the trivial formula

$$h(1) = h(0) + \int_0^1 \frac{\mathrm{d}h}{\mathrm{d}t} \,\mathrm{d}t,$$

$$\bar{f}(p^1, \ldots, p^n) = \bar{f}(0, \ldots, 0) + \int_0^1 \sum_{i=1}^{n} \frac{\partial \bar{f}(tp^1, \ldots, tp^n)}{\partial x^i} p^i \,\mathrm{d}t$$

$$= \bar{f}(0, \ldots, 0) + \sum_{i=1}^{n} p^i \int_0^1 \frac{\partial \bar{f}(tp^1, \ldots, tp^n)}{\partial x^i} \,\mathrm{d}t .$$

Set

$$\tilde{f}_i(p^1, \ldots, p^n) = \int_0^1 \frac{\partial \tilde{f}(tp^1, \ldots, tp^n)}{\partial x^i} \, \mathrm{d}t .$$

Thus we get (2.4); it remains to show that $\tilde{f}_i$ are of class C^∞ on V_1. For this, we have to use the following well known assertion: Let $h(p^1, \ldots, p^n, t)$ be a function of class C^1 on $\mathbf{R}^{n+1}$ and $k(p^1, \ldots, p^n) = \int_0^1 h(p^1, \ldots, p^n, t) \, \mathrm{d}t$, then k is of class C^1 and

$$\frac{\partial k(p^1, \ldots, p^n)}{\partial x^i} = \int_0^1 \frac{\partial h(p^1, \ldots, p^n, t)}{\partial x^i} \, \mathrm{d}t .$$

Applying this to the functions of the type

$$h(p^1, \ldots, p^n, t) = \frac{\partial^m \tilde{f}(tp^1, \ldots, tp^n)}{\partial x^{i_1} \ldots \partial x^{i_m}} ,$$

we get the desired property. QED.

The proof of the following result is easy.

Theorem 2.3. *Let M, N be manifolds, $\varphi\colon M \to N$ a mapping and $m \in M$. To each $t \in T_m(M)$, let us associate the mapping $\varphi_*(t)\colon \mathscr{F}^\infty(\varphi(m), N) \to \mathbf{R}$ by*

(2.5) $\quad \varphi_*(t)(f) = t(f \circ \varphi) \quad$ *for* $\quad f \in \mathscr{F}^\infty(\varphi(m), N)$.

Then $\varphi_*(t) \in T_{\varphi(m)}(N)$ and $\varphi_*\colon T_m(M) \to T_{\varphi(m)}(N)$ is a homomorphism.

Definition 2.3. *The homomorphism φ_* of Theorem 2.3 is called the* differential *of φ at m.*

There are various notations for the differential: $\varphi_* = \varphi_*(m) = (\mathrm{d}\varphi)_m = \varphi_m'$.
The following is also very easy to prove.

Theorem 2.4. *Let M, N, P be manifolds, $\varphi\colon M \to N$ and $\psi\colon N \to P$ mappings, $m \in M$. Then the diagram*

$$\begin{array}{ccc} T_m(M) & \xrightarrow{(\psi \circ \varphi)_*} & T_{\psi(\varphi(x))}(P) \\ & \searrow^{\varphi_*} \quad \nearrow^{\psi_*} & \\ & T_{\varphi(m)}(N) & \end{array}$$

is commutative.

The purpose of the following theorem is to express φ_* in terms of local coordinates.

Theorem 2.5. *Let M and N be manifolds, $\mu\colon U \to \mathbf{R}^n$ and $\nu\colon V \to \mathbf{R}^m$ maps of M and N resp., $\varphi\colon M \to N$ a mapping such that $\varphi(U) \subset V$, $m \in M$. The mapping $F = \nu \circ \varphi \circ \mu^{-1}\colon \mu(U) \to \mathbf{R}^m$ be given by*

(2.6) $\quad y^\alpha = F^\alpha(x^1, \ldots, x^n); \quad \alpha = 1, \ldots, m;$

let $\mu(m) = (\overset{\circ}{x}{}^1, \ldots, \overset{\circ}{x}{}^n) \in \mathbf{R}^n$. *Then*

$$\varphi^* \left(\left. \frac{\partial}{\partial x^i} \right|_m \right) = \sum_{\alpha=1}^{m} \frac{\partial F^\alpha(\overset{\circ}{x}{}^1, \ldots, \overset{\circ}{x}{}^n)}{\partial x^i} \cdot \left. \frac{\partial}{\partial y^\alpha} \right|_{\varphi(m)} \quad \textit{for} \quad i = 1, \ldots, n. \tag{2.7}$$

Proof. Let $f \in \mathscr{F}^\infty(\varphi(m), N)$. According to (2.5) and (2.2),

$$\varphi_* \left(\left. \frac{\partial}{\partial x^i} \right|_m \right) (f) = \left. \frac{\partial}{\partial x^i} \right|_m (f \circ \varphi) = \left. \frac{\partial (f \circ \varphi \circ \mu^{-1})}{\partial x^i} \right|_{\mu(m)}$$

$$= \left. \frac{\partial (f \circ \nu^{-1} \circ \nu \circ \varphi \circ \mu^{-1})}{\partial x^i} \right|_{\mu(m)} = \left. \frac{\partial (f \circ \nu^{-1} \circ F)}{\partial x^i} \right|_{\mu(m)}.$$

Further,

$$\left. \frac{\partial}{\partial y^\alpha} \right|_{\varphi(m)} (f) = \left. \frac{\partial (f \circ \nu^{-1})}{\partial y^\alpha} \right|_{(\nu \circ \varphi)(m)} = \left. \frac{\partial (f \circ \nu^{-1})}{\partial y^\alpha} \right|_{(\nu \circ \varphi \circ \mu^{-1} \circ \mu)(m)} = \left. \frac{\partial (f \circ \nu^{-1})}{\partial y^\alpha} \right|_{F(\mu(m))}.$$

The chain rule implies

$$\left. \frac{\partial (f \circ \nu^{-1} \circ F)}{\partial x^i} \right|_{\mu(m)} = \sum_{\alpha=1}^{m} \frac{\partial F^\alpha(\mu(m))}{\partial x^i} \left. \frac{\partial (f \circ \nu^{-1})}{\partial y^\alpha} \right|_{F(\mu(m))}$$

for arbitrary f, i.e., we get (2.7). QED.

Now, let us consider the manifold $\mathbf{R}$, its atlas consisting of the map id: $\mathbf{R} \to \mathbf{R}$. Let $r \in \mathbf{R}$. Then $\dim T_r(\mathbf{R}) = 1$, and

$$\left. \frac{\partial}{\partial x} \right|_r : \mathscr{F}^\infty(r, \mathbf{R}) \to \mathbf{R}, \quad \left. \frac{\partial}{\partial x} \right|_r (f) = \frac{\mathrm{d}f(r)}{\mathrm{d}x},$$

is the basis of $T_r(\mathbf{R})$. Thus each vector $t \in T_r(\mathbf{R})$ may be written as $t = a \cdot \partial/\partial x|_r$, $a \in \mathbf{R}$. The mapping

$$\iota_r : T_r(\mathbf{R}) \to \mathbf{R}, \quad \iota_r \left(a \cdot \left. \frac{\partial}{\partial x} \right|_r \right) = a, \tag{2.8}$$

is an isomorphism. Let M be a manifold, $\varphi: M \to \mathbf{R}$ a function, $m \in M$. Then we have the homomorphism $\varphi_* : T_m(M) \to T_{\varphi(m)}(\mathbf{R})$.

Definition 2.4. *Let M be a manifold, $\varphi: M \to \mathbf{R}$ a function, $m \in M$. The homomorphism*

$$(\mathrm{d}\varphi)_m : T_m(M) \to \mathbf{R}, \qquad (\mathrm{d}\varphi)_m = \iota_{\varphi(m)} \circ \varphi_*, \tag{2.9}$$

is called the differential *of f at m.*

3. Exterior differentiation

Definition 3.1. *Let M be a manifold and $U \subset M$ an open subset. A* tangent vector field *on U is a mapping $t: U \to \bigcup_{m \in U} T_m(M)$ such that $t_m \equiv t(m) \in T_m(M)$ for each $m \in U$.* The tangent vector field t on U is called *differentiable* (i.e., of class C^∞) if, for each function $f: U \to \mathbf{R}$, the function tf, defined by $(tf)(m) = t_m(f)$, is of class C^∞.

The set U being a coordinate neighbourhood with local coordinates $(x^1, \ldots, x^n)$, the tangent vector field t may be written as $t = a^i(x) \cdot \partial/\partial x^i$. Then t is differentiable if and only if the functions $a^i(x)$ are of class C^∞. We are going to consider the differentiable tangent vector fields only.

Theorem 3.1. *Let M be a manifold, $U \subset M$ an open subset and t, t', t'' tangent vector fields on U. For each $m \in M$, define the mapping $[t, t']_m : \mathscr{F}^\infty(m, M) \to \mathbf{R}$ by*

$$[t, t']_m\,(f) = t_m(t'f) - t_m'(tf). \tag{3.1}$$

Then $[t, t']_m \in T_m(M)$ and $[t, t']$ is a vector field on U. Analogously for $[t, t'']$ and $[t', t'']$. We have

$$[rt + r't', t''] = r[t, t''] + r'[t', t''], \tag{3.2}$$

$$[t, r't' + r''t''] = r'[t, t'] + r''[t, t''] \quad \textit{for} \quad r, r', r'' \in \mathbf{R};$$

$$[t, t'] = -[t', t]; \tag{3.3}$$

$$\big[[t, t'], t''\big] + \big[[t', t''], t\big] + \big[[t'', t], t'\big] = 0. \tag{3.4}$$

Proof. Indeed,

$$[t, t']_m\,(f + g) = [t, t']_m\,(f) + [t, t']_m\,(g),$$

$$\begin{aligned}[t, t']_m\,(fg) &= t_m(t'f \cdot g + f \cdot t'g) - t_m'(tf \cdot g + f \cdot tg)\\ &= t_m(t'f) \cdot g(m) + t_m'f \cdot t_m g + t_m f \cdot t_m' g + f(m) \cdot t_m(t'g)\\ &\quad - t_m'(tf) \cdot g(m) - t_m f \cdot t_m' g - t_m f \cdot t_m' g - f(m) \cdot t_m'(tg)\\ &= [t, t']_m\,(f) \cdot g(m) + f(m) \cdot [t, t']_m\,(g),\end{aligned}$$

i.e., $[t, t']_m \in T_m(M)$. (3.2) and (3.3) are trivial. Applying the left-hand side of (3.4) to an arbitrary function $f : U \to \mathbf{R}$, we get

$$\begin{aligned}&[t, t']\,(t''f) - t''([t, t']\,f) + [t', t'']\,(tf)\\ &\quad - t([t', t'']\,f) + [t'', t]\,(t'f) - t'([t'', t]\,f)\\ &= (tt't''f - t'tt''f) - (t''tt'f - t''t'tf) + (t't''tf - t''t'tf)\\ &\quad - (tt't''f - tt''t'f) + (t''tt'f - tt''t'f) - (t't''tf - t'tt''f) = 0.\end{aligned}$$

QED.

The proof of the following theorem is trivial.

Theorem 3.2. *Let M, N be manifolds, $\varphi : M \to N$ a mapping and t, t' tangent vector fields on $U \subset M$. Then, $\varphi_* t$ and $\varphi_* t'$ being vector fields on $\varphi(U)$,*

$$[\varphi_* t, \varphi_* t'] = \varphi_*[t, t'] \quad \text{on} \quad \varphi(U) \subset N. \tag{3.5}$$

Definition 3.2. *Let M be a manifold and $U \subset M$ an open subset. An* (exterior) r-form *on U is a mapping $\omega : U \to \bigcup_{m \in U} \wedge^r T_m{}^*(M)$ such that $\omega_m \equiv \omega(m) \in \wedge^r T_m{}^*(M)$. The r-form ω is called* differentiable (*i.e.*, of class C^∞) *if, for each tangent vector fields $t_1, \ldots, t_r$ on U, $\omega(t_1 \wedge \cdots \wedge t_r)$ is a function of class C^∞.*

Let the set U be a coordinate neighbourhood with local coordinates $(x^1, \ldots, x^n)$. The space $T_m(M)$, $m \in U$, has the natural basis $\partial/\partial x^i|_m$. Denote by $\mathrm{d}x^i|_m$ the dual basis of $T_m{}^*(M)$. Each r-form ω may then be written as

$$(3.6) \qquad \omega = \frac{1}{r!} \sum_{i_1=1}^{n} \cdots \sum_{i_r=1}^{n} b_{i_1 \ldots i_r}(x)\, \mathrm{d}x^{i_1} \wedge \cdots \wedge \mathrm{d}x^{i_r}, \quad b_{i_1 \ldots i_r} \text{ antisymmetric.}$$

Evidently, ω is differentiable if and only if $b_{i_1 \ldots i_r}$ are differentiable functions.

Definition 3.2. *Be given a manifold M. An* exterior differentiation *is an operator* d, *which associates to each r-form ω on $U \subset M$ an $(r+1)$-form $\mathrm{d}\omega$ on U and has the following properties (all forms are to be defined on U, U being an arbitrary open subset of M)*: (i)

$$(3.7) \qquad \mathrm{d}(r_1\omega_1 + r_2\omega_2) = r_1\, \mathrm{d}\omega_1 + r_2\, \mathrm{d}\omega_2$$

for r-forms ω_1, ω_2 and $r_1, r_2 \in \mathbf{R}$; (ii) *for a 0-form f (i.e., a function $f \colon U \to \mathbf{R}$), $\mathrm{d}f$ is the differential of f (see* Definition 2.4); (iii) *for an r-form ω_1 and an s-form ω_2,*

$$(3.8) \qquad \mathrm{d}(\omega_1 \wedge \omega_2) = \mathrm{d}\omega_1 \wedge \omega_2 + (-1)^r\, \omega_1 \wedge \mathrm{d}\omega_2;$$

(iv) *we have* $\mathrm{d}^2 = 0$, *i.e., for an r-form ω,*

$$(3.9) \qquad \mathrm{d}(\mathrm{d}\omega) = 0.$$

Theorem 3.3. *The exterior differentiation is a local operator in the following sense: Let $U \subset U_1 \subset X$ be open sets and let ω_1, ω_2 be r-forms defined on U_1. If $\omega_1|_U = \omega_2|_U$, we have $\mathrm{d}\omega_1|_U = \mathrm{d}\omega_2|_U$.*

Proof. Let $\omega = \omega_1 - \omega_2$, i.e., $\omega|_U = 0$. Choose $m \in U$. It is well known that the point m has a neighbourhood V such that $\bar{V} \subset U$ and there is a function $\varrho \colon M \to \mathbf{R}$ with $\varrho(V) = 0$ and $\varrho(M - U) = 1$. Then $\omega = \varrho\omega$, and, according to (3.8), $\mathrm{d}\omega = \mathrm{d}\varrho \wedge \omega + \varrho\, \mathrm{d}\omega$, $(\mathrm{d}\omega)_m = (\mathrm{d}\varrho)_m \wedge \omega_m + \varrho(m)\, (\mathrm{d}\omega)_m$. From this, $(\mathrm{d}\omega)_m = 0$. QED.

Theorem 3.4. *There exists exactly one operator of the exterior differentiation.*

Proof. Let there exist two such operators $\mathrm{d}_1, \mathrm{d}_2$ which are different. This implies the existence of an r-form ω on U such that $(\mathrm{d}_1\omega)_m \neq (\mathrm{d}_2\omega)_m$ for some point $m \in U$. Let us choose a coordinate neighbourhood U_1 of $_m$; let ω be expressed, in U_1, by (3.6). From the properties of the exterior differentiation,

$$(3.10) \qquad \mathrm{d}\omega = \frac{1}{r!} \sum_{i_1=1}^{n} \cdots \sum_{i_r=1}^{n} \mathrm{d}a_{i_1 \ldots i_r}(x) \wedge \mathrm{d}x^{i_1} \wedge \cdots \wedge \mathrm{d}x^{i_r}$$
$$= \frac{1}{r!} \sum_{i=1}^{n} \sum_{i_1=1}^{n} \cdots \sum_{i_r=1}^{n} \frac{\partial a_{i_1 \ldots i_r}}{\partial x^i}\, \mathrm{d}x^i \wedge \mathrm{d}x^{i_1} \wedge \cdots \wedge \mathrm{d}x^{i_r}$$

for each exterior differentiation d. Thus $(\mathrm{d}_1\omega)_m = (\mathrm{d}_2\omega)_m$, a contradiction. The formula (3.10) exhibits the existence of d: in each coordinate neighbourhood, define $\mathrm{d}\omega$ by (3.10). I leave to the reader to show that these formulas coincide on the intersection of two coordinate neighbourhoods and d is really an exterior differentiation. QED.

Theorem 3.5. *Let M be a manifold, $U \subset M$ an open subset, ω a 1-form on U, t and t' tangent vector fields on U. Then*

$$(3.11) \qquad (\mathrm{d}\omega)(t, t') = t\omega(t') - t'\omega(t) - \omega([t, t']).$$

Proof. It is sufficient to prove (3.11) in any coordinate neighbourhood $U_1 \subset U$. The formula (3.11) being valid for ω_1 and ω_2, it is valid for $\omega_1 + \omega_2$ as well. Now, suppose its validity for a 1-form on U_1, and let $f\colon U_1 \to \mathbf{R}$ be a function. Then

$$\begin{aligned} \mathrm{d}(f\omega)(t, t') &= (\mathrm{d}f \wedge \omega + f\,\mathrm{d}\omega)(t, t') \\ &= \begin{vmatrix} \mathrm{d}f(t) & \mathrm{d}f(t') \\ \omega(t) & \omega(t') \end{vmatrix} + f \cdot \mathrm{d}\omega(t, t') \\ &= tf \cdot \omega(t') - \omega(t) \cdot t'f + f \cdot t\omega(t') - f \cdot t'\omega(t) - f\omega([t, t']) \\ &= t\big(f\omega(t')\big) - t'\big(f\omega(t)\big) - f\omega([t, t']), \end{aligned}$$

i.e., (3.11) is valid for $f\omega$ as well. Thus we may restrict our proof to the form $\omega = \mathrm{d}x^k$ and, because of the bilinearity of $\mathrm{d}\omega$, to the vector fields $t = \partial/\partial x^i$, $t' = \partial/\partial x^j$. Then $\mathrm{d}\omega = 0$, $[t, t'] = 0$, and the right-hand side of (3.11) is equal to

$$\frac{\partial}{\partial x^i}(\delta_j^k) - \frac{\partial}{\partial x^j}(\delta_i^k) = 0.$$

QED.

Analogously, we get the general

Theorem 3.6. *Let M be a manifold, $U \subset M$ an open subset, ω an r-form on U, $t_1, \ldots, t_{r+1}$ tangent vector fields on U. Then*

$$(3.12) \qquad \begin{aligned} \mathrm{d}\omega(t_1, \ldots, t_{r+1}) &= \sum_{i=1}^{r+1} (-1)^{i-1}\, t_i\omega(t_1, \ldots, t_{i-1}, t_{i+1}, \ldots, t_{r+1}) \\ &\quad + \sum_{1 \leq i < j \leq r+1} (-1)^{i+j}\, \omega([t_i, t_j], t_1, \ldots, t_{i-1}, t_{i+1}, \ldots, t_{j-1}, t_{j+1}, \ldots, t_{r+1}). \end{aligned}$$

Let M and N be manifolds and $\varphi\colon M \to N$ a mapping. For an r-form ω on $V \subset N$ define the r-form $\varphi^*\omega$ on $U = \varphi^{-1}(V) \subset M$ by means of the formula

$$(3.13) \qquad \varphi^*\omega(t_1, \ldots, t_r) = \omega(\varphi_* t_1, \ldots, \varphi_* t_r)$$

for tangent vector fields $t_1, \ldots, t_r$ on U. The following theorem is evident because of (3.12).

Theorem 3.7. *Let M, N be manifolds and $\varphi\colon M \to N$ a mapping. Then*

$$(3.14) \qquad \varphi^* \circ \mathrm{d} = \mathrm{d} \circ \varphi^*$$

in the following sense: ω being defined on V, we have $\mathrm{d}(\varphi^*\omega) = \varphi^*(\mathrm{d}\omega)$ on $U = \varphi^{-1}(V)$.

Theorem 3.8 (Poincaré lemma). *Let M be a manifold, $U \subset M$ an open subset, ω an r-form on U ($r \geqq 1$), $m \in U$. If $\mathrm{d}\omega = 0$ on U, there is a neighbourhood $U_1 \subset U$ of m and an $(r-1)$-form ω_1 on U_1 such that $\mathrm{d}\omega_1 = \omega$ on U_1.*

Proof. Let us suppose that U_1 is a coordinate neighbourhood $\mu_1\colon U_1 \to \mathbf{R}^n$ with $\mu_1(m) = (0, \ldots, 0)$ and μ_1 mapping U_1 onto a ball centered in the origin of $\mathbf{R}^n$. Let us write, in U_1,

$$\omega = \sum_{1 \leqq i_1 < \cdots < i_r \leqq n} a_{i_1 \ldots i_r}(x)\, \mathrm{d}x^{i_1} \wedge \cdots \wedge \mathrm{d}x^{i_r}.$$

The $(r - 1)$-form

$$k\omega = \sum_{1 \leqq i_1 < \cdots < i_{r-1} \leqq n} (ka)_{i_1 \ldots i_{r-1}}\, \mathrm{d}x^{i_1} \wedge \cdots \wedge \mathrm{d}x^{i_{r-1}}$$

be defined by

$$(ka)_{i_1 \ldots i_{r-1}}(x) = \int_0^1 \sum_{j=1}^{n} t^{r-1} x^j a_{j i_1 \ldots i_{r-1}}(tx^1, \ldots, tx^n)\, \mathrm{d}t.$$

By a direct calculation, we get

$$k(\mathrm{d}\omega) + \mathrm{d}(k\omega) = \omega \quad \text{for an } r\text{-form with } r > 0,$$

$$k(\mathrm{d}f) = f - f(m) \quad \text{for a function } f.$$

The proof follows: take $\omega_1 = k\omega$. QED.

III. Methods of global differential geometry

1. Stokes' theorem

By a *standard q-simplex* σ_q we shall mean a subset of $\mathbf{R}^q$ consisting of points $(t^1, \ldots, t^q)$ satisfying

$$(1.1) \qquad t^1 \geqq 0, \ldots, t^q \geqq 0, \quad t^1 + \cdots + t^q \leqq 1.$$

The *vertices of* σ_q *are then the points*

$$E_0 = (0, \ldots, 0), \quad E_1 = (1, 0, \ldots, 0), \ \ldots, \quad E_q = (0, \ldots, 0, 1).$$

The *standard* 0*-simplex* σ_0 is, by definition, the point E_0. The *barycentric coordinates* of the point $(t^1, \ldots, t^q) \in \sigma_q$ are $(t^0, t^1, \ldots, t^q)$ with $t^0 = 1 - t^1 - \cdots - t^q$. The i-th *face* $\sigma_q{}^i$; $i = 0, \ldots, q$; of σ_q is the subset of its points satisfying $t^i = 0$.

Let $\sigma_{q-1} \subset \mathbf{R}^{q-1}$ be the standard $(q-1)$-simplex. The mapping

$$(1.2) \qquad \tilde{\sigma}^j_{q-1} \colon \sigma_{q-1} \to \mathbf{R}^q; \quad j = 0, \ldots, q;$$

be defined as follows:

$$(1.3) \qquad \tilde{\sigma}^0_{q-1}(\tau^1, \ldots, \tau^{q-1}) = (1 - \tau^1 - \cdots - \tau^{q-1}, \tau^1, \ldots, \tau^{q-1}),$$

$$\tilde{\sigma}^j_{q-1}(\tau^1, \ldots, \tau^{q-1}) = (\tau^1, \ldots, \tau^{j-1}, 0, \tau^j, \ldots, \tau^{q-1}) \quad \text{for} \quad j = 1, \ldots, q.$$

Obviously, $\tilde{\sigma}^j_{q-1}$ maps σ_{q-1} onto the j-th face of $\sigma_q \subset \mathbf{R}^q$.

Definition 1.1. *Let M be a differentiable manifold. A* singular q-simplex *in M is a mapping $s_q \colon \sigma_q \to M$ such that there is an open set $U \subset \mathbf{R}^q$ and a (differentiable) mapping $s' \colon U \to M$ such that $s_q \subset U$ and $s'|_{\sigma_q} = s_q$. A* singular q-chain *c_q in M is a real-valued function on the set of all singular q-simplexes of M taking non-zero values just on a finite number of them.*

The set of singular q-chains c_q in M is made, in the obvious manner, into a vector space over $\mathbf{R}$; let us denote it by $C_q(M)$.

Definition 1.2. *Let $s_q \colon \sigma_q \to M$ be a singular q-simplex in M. Its* boundary *is defined as the singular $(q-1)$-chain*

$$(1.4) \qquad \partial s_q = \sum_{j=1}^{q} (-1)^j \, (s_q \circ \tilde{\sigma}^j_{q-1}).$$

The boundary operator

$$(1.5) \qquad \partial \colon C_q(M) \to C_{q-1}(M)$$

be defined by

$$(1.6)\qquad \partial\Big(\sum_i a_i \cdot s_q^{(i)}\Big) = \sum_i a_i \cdot \partial s_q^{(i)}; \qquad a_i \in \mathbf{R}.$$

The q-chain c_q is called the q-cycle *if $q = 0$ or $\partial c_q = 0$; c_q is called the q*-boundary *if there is a $(q+1)$-chain c_{q+1} such that $c_q = \partial c_{q+1}$.*

The following theorem is easy to prove by a direct calculation.

Theorem 1.1. *We have*

$$(1.7)\qquad \partial^2 = 0,$$

i.e., for each q-chain c_q, $\partial(\partial c_q) = 0$, i.e., ∂c_q is always a $(q-1)$-cycle.

Each exterior q-form on $\mathbf{R}^q$ (or a part of it) may be written as

$$(1.8)\qquad \varphi = f(t^1, \ldots, t^q)\, \mathrm{d}t^1 \wedge \cdots \wedge \mathrm{d}t^q.$$

Definition 1.3. *Let the form* (1.8) *be defined in some neighbourhood of the standard q-simplex σ_q. Then $\int_{\sigma_q} \varphi$ be defined by*

$$(1.9)\qquad \int_{\sigma_q} \varphi = \int_{\sigma_q} f(t^1, \ldots, t^q)\, \mathrm{d}t^1 \cdots \mathrm{d}t^q$$

$$= \int_{t^q=0}^{1} \int_{t^{q-1}=0}^{1-t^q} \cdots \int_{t^1=0}^{1-t^q-\cdots-t^2} f(t^1, \ldots, t^q)\, \mathrm{d}t^1 \cdots \mathrm{d}t^q.$$

Let $s_q\colon \sigma_q \to M$ be a singular q-simplex in M, and let φ be a q-form defined in some neighbourhood of $s_q(\sigma_q) \subset M$. For $q > 0$, set

$$(1.10)\qquad \int_{s_q} \varphi = \int_{\sigma_q} s_q^*(\varphi);$$

for $q = 0$, set

$$(1.11)\qquad \int_{s_0} \varphi = \varphi(s_0).$$

*Let $c_q = \sum_i a_i s_q^{(i)}$ be a q-chain in M, and let φ be a q-form defined in some neighbour*hood of $\bigcup_i s_q^{(i)}(\sigma_q)$. *Then set*

$$(1.12)\qquad \int_{c_q} \varphi = \sum_i a_i \int_{s_q^{(i)}} \varphi.$$

Theorem 1.2 (Stokes' theorem). *Let c_q be a q-chain in M and φ a $(q-1)$-form on M* (it is sufficient to suppose that φ is defined just on a neighbourhood of the simplexes appearing in c_q). *Then*

$$(1.13)\qquad \int_{c_q} \mathrm{d}\varphi = \int_{\partial c_q} \varphi.$$

Proof. First of all, let us prove our theorem for a standard q-simplex $\sigma_q \subset \mathbf{R}^q = M$. Let $q = 1$. Then φ is a real-valued function defined on some neighbourhood of $\sigma_1 = \langle 0, 1\rangle$. σ_0 being the standard 0-simplex, we have $\sigma_0^1(\sigma_0) = (0)$, $\sigma_0^0(\sigma_0) = (1)$, i.e., $\partial\sigma_1 = (1) - (0)$. Then

$$\int_{\sigma_1} \mathrm{d}\varphi(x) = \int_0^1 \frac{\mathrm{d}\varphi(x)}{\mathrm{d}x}\,\mathrm{d}x = \varphi(1) - \varphi(0), \quad \int_{\partial\sigma_1} \varphi(x) = \int_{(1)-(0)} \varphi(x) = \varphi(1) - \varphi(0),$$

and the Stokes theorem holds true. For $q \geqq 2$, we may write

$$\varphi = \sum_{i=1}^{q} \varphi_{(i)},$$

where

$$\varphi_{(i)} = \varphi_i(t^1, \ldots, t^q)\,\mathrm{d}t^1 \wedge \cdots \wedge \mathrm{d}t^{i-1} \wedge \mathrm{d}t^{i+1} \wedge \cdots \wedge \mathrm{d}t^q.$$

Both sides of (1.12) depending linearly on φ, it is sufficient to prove (1.12) just for the form $\varphi_{(i)}$. Because of the possible changes of variables $t^1, \ldots, t^q$, it is sufficient to prove it just for the form

$$\psi = p(t^1, \ldots, t^q)\,\mathrm{d}t^2 \wedge \cdots \wedge \mathrm{d}t^q.$$

Of course,

$$\mathrm{d}\psi = \frac{\partial p}{\partial t^1}\,\mathrm{d}t^1 \wedge \mathrm{d}t^2 \wedge \cdots \wedge \mathrm{d}t^q,$$

and

$$(1.14) \quad \int_{\sigma_q} \mathrm{d}\psi = \int_{\sigma_q} \frac{\partial p}{\partial t^1}\,\mathrm{d}t^1\,\mathrm{d}t^2 \cdots \mathrm{d}t^q = \int_{t^q=0}^{1} \int_{t^{q-1}=0}^{1-t^q} \cdots \int_{t^1=0}^{1-t^q-\cdots-t^2} \frac{\partial p}{\partial t^1}\,\mathrm{d}t^1 \cdots \mathrm{d}t^q$$

$$= \int_{t^q=0}^{1} \int_{t^{q-1}=0}^{1-t^q} \cdots \int_{t^2=0}^{1-t^q-\cdots-t^3} \{p(1 - t^q - \cdots - t^2, t^2, \ldots, t^q)$$

$$- p(0, t^2, \ldots, t^q)\}\,\mathrm{d}t^2 \cdots \mathrm{d}t^q.$$

Further, it is easy to see that

$$(\sigma^1_{q-1})^*\,\psi = p(0, \tau^1, \tau^2, \ldots, \tau^{q-1})\,\mathrm{d}\tau^1 \wedge \cdots \wedge \mathrm{d}\tau^{q-1},$$

$$(\sigma^j_{q-1})^*\,\psi = 0 \quad \text{for} \quad j \geqq 2,$$

$$(\sigma^0_{q-1})^*\,\psi = p(1 - \tau^1 - \cdots - \tau^{q-1}, \tau^1, \ldots, \tau^{q-1})\,\mathrm{d}\tau^1 \wedge \cdots \wedge \mathrm{d}\tau^{q-1}.$$

Thus

$$(1.15) \quad \int_{\partial\sigma_q} \psi = \sum_{i=0}^{q} (-1)^i \int_{\sigma_{q-1}} (\sigma^i_{q-1})^*\,\psi;$$

the right-hand term of (1.15) being equal to the last term of (1.14), we are done.

Now, let $s_q\colon \sigma_q \to M$ be a singular q-simplex. Using Theorem II.3.7, we get

$$\int_{s_q} \mathrm{d}\varphi = \int_{\sigma_q} s_q{}^*(\mathrm{d}\varphi) = \int_{\sigma_q} \mathrm{d}(s_q{}^*\varphi) = \int_{\partial\sigma_q} s_q{}^*\varphi = \int_{\partial\sigma_q} \varphi.$$

QED.

2. Maximum principle

In a domain $G \subset \mathbf{R}^n$, let us consider a second order differential operator of the form

$$(2.1) \qquad L = \sum_{i,j=1}^{n} a_{ij}(x) \frac{\partial^2}{\partial x^i \, \partial x^j} + \sum_{i=1}^{n} a_i(x) \frac{\partial}{\partial x^i}, \quad a_{ij}(x) = a_{ji}(x);$$

L is said to be *elliptic* in G if the quadratic form $\sum_{i,j=1}^{n} a_{ij}(x)\, \xi^i \xi^j$ is positive definite for each $(x) \in G$.

Theorem 2.1 (E. Hopf). *Let* $B^n = \{(x^1, \ldots, x^n) \in \mathbf{R}^n;\ (x^1)^2 + \cdots + (x^n)^2 < 1\}$ *be the open unit ball centered at the origin of* $\mathbf{R}^n$. *Suppose that the function* $f\colon B^n \to \mathbf{R}$ *satisfies:* (i) $Lf \geqq 0$ *in* B^n; (ii) *there is a point* $p_0 \in B^n$ *such that* $f(p) \leqq f(p_0)$ *for each* $p \in B^n$. *Then* $f(p) = f(p_0)$ *for each* $p \in B^n$. *In the conditions* (i) *and* (ii), *we may replace* $Lf \geqq 0$, $f(p) \leqq f(p_0)$ *by* $Lf \leqq 0$ *and* $f(p) \geqq f(p_0)$ *resp.*

Proof. Write $f_0 = f(p_0)$, and let us suppose $f(p) < f_0$ for some point $p \in B^n$. Obviously, the point p with this property may be chosen in such a way that

$$d(p, F) = \inf_{x \in F} d(p, x) < d(p, \partial B^n) = \inf_{y \in \partial B^n} d(p, y),$$

F being the set $\{(x) \in B^n;\ f(x) = f_0\}$, ∂B^n the boundary of B^n and d the usual Euclidean distance in $\mathbf{R}^n$. There exists a ball B_1 centered at p and such that $f(q) < f_0$ for each $q \in B_1$. Let us increase the radius of this ball B_1; once we get a ball B_2 such that there is a point $p_1 \in B^n \cap \partial B_2$ and we have $f(q) < f_0$ for the points $q \in B^n \cap B_2$. Take another ball B_3 with the following properties: $p_1 \in \partial B_3$, ∂B_3 and ∂B_2 have common tangent hyperplane at p_1, $B_3 \subset B_2 \cap B^n$. Let R be the radius of B_3. Of course, $f(p_1) = f_0$ and $f(q) < f_0$ for each $q \in B_3 \cup \partial B_3 - \{p_1\}$. Finally, let B_4 be a ball centered at p_1, $B_4 \subset B^n$, and let $R' < R$, R' being the radius of B_4. Write $F_i = \partial B_4 \cap (B_3 \cup \partial B_3)$, $F_0 = \partial B_4 - F_i$. Obviously, $f(q) < f_0$ for $q \in F_i$. Thus there is an $\varepsilon > 0$ such that

$$(2.2) \qquad f(q) \leqq f_0 - \varepsilon \quad \text{for} \quad q \in F_i, \qquad f(q) \leqq f_0 \quad \text{for} \quad q \in F_0.$$

In $\mathbf{R}^n$, let us take new coordinates $y^i = x^i - \mathring{x}^i$, $(\mathring{x}^1, \ldots, \mathring{x}^n)$ being the center of the ball B_3; we get

$$(2.3) \qquad L = \sum_{i,j=1}^{n} a_{ij}(y + \mathring{x}) \frac{\partial^2}{\partial y^i \, \partial y^j} + \sum_{i=1}^{n} a_i(y + \mathring{x}) \frac{\partial}{\partial y^i}.$$

Consider the function $\psi\colon \mathbf{R}^n \to \mathbf{R}$ defined by

$$(2.4) \qquad \psi(x) = \mathrm{e}^{-\alpha r^2} - \mathrm{e}^{-\alpha R^2},$$

$\alpha > 0$ being a constant and $r^2 = (y^1)^2 + \cdots + (y^2)^2$. It is easy to see that

$$(2.5) \qquad L\psi = \mathrm{e}^{-\alpha r^2} \left\{ 4\alpha^2 \sum_{i,j=1}^{n} a_{ij} y^i y^j - 2\alpha \sum_{i=1}^{n} (a_i y^i + a_{ii}) \right\}.$$

Because of $R' < R$, the center of the ball B_3 (i.e., the origin of the new coordinate system $y^1, \ldots, y^n$) is situated outside of $B_4 \cup \partial B_4$. On $B_4 \cup \partial B_4$, we have $\sum_{i,j=1}^{n} a_{ij} x^i x^j > 0$

i.e., $\sum_{i,j=1}^{n} a_{ij}x^i x^j \geqq \text{const} > 0$. The constant α being taken large enough, we may attain $L\psi > 0$ on B_4. On the other hand,

$$\psi(q) < 0 \quad \text{for} \quad q \in F_0, \qquad \psi(p_1) = 0. \tag{2.6}$$

Define another function $\Phi: B^n \to \mathbf{R}$ by $\Phi(x) = f(x) + \delta\psi(x)$, $\delta > 0$ being a constant. This constant δ be chosen in such a way that $\Phi(q) < f_0$ for each $q \in F_i$; such a choise is possible because of (2.2_1). From (2.2_2) and (2.6_1), $\Phi(q) < f_0$ for each $q \in F_0$, and we have $\Phi(q) < f_0$ for each $q \in \partial B_1 = F_0 \cup F_i$. Further, $\Phi(p_1) = f_0$ because of (2.6_2). Thus the function Φ attains its maximum in $B_4 \cup \partial B_4$ inside of B_4; suppose that

$$\Phi(q_0) = \max_{q \in B_4 \cup \partial B_4} \Phi(q), \qquad q_0 \in B_4.$$

On B_4, we have $L\varphi \geqq 0$ and $L\psi > 0$, i.e., $L\Phi > 0$. On the other hand,

$$L\Phi(q_0) = \sum_{i,j=1}^{n} a_{ij}(q_0) \frac{\partial^2 \Phi(q_0)}{\partial y^i \, \partial y^j}$$

and we should have

$$\sum_{i,j=1}^{n} \frac{\partial^2 \Phi(q_0)}{\partial y^i \, \partial y^j} \lambda^i \lambda^j \leqq 0 \quad \text{for each} \quad \lambda^1, \ldots, \lambda^n \in \mathbf{R}. \tag{2.7}$$

The form $\sum_{i,j=1}^{n} a_{ij}\xi^i\xi^j$ being positive definite and the left-hand side form in (2.7) negative indefinite, we have $L\Phi(q_0) \leqq 0$, this being a contradiction to $L\Phi > 0$ on B_4. QED.

Theorem 2.1 holds true if replacing B^n by a domain $U \subset \mathbf{R}^n$. Indeed, let $p \in U$, and let us connect p_0 and p by an arc $\gamma \subset U$. Then there are balls $B_i \subset \mathbf{R}^n$ such that $\gamma \subset \bigcup_i B_i \subset U$, and it is easy to see that f is equal to $f(p_0)$ in $\bigcup_i B_i$.

Let $U \subset \mathbf{R}^n$ be a domain, and let $F: U \to V = F(U) \subset \mathbf{R}^n$ be a mapping. On U, be given an operator (2.1). Let us consider the operator F_*L on V defined by

$$(F_*L)\,\tilde{f} = L(\tilde{f} \circ F) \quad \text{for each function } \tilde{f}: V \to \mathbf{R}. \tag{2.8}$$

Let F be expressed by $y^i = F^i(x^1, \ldots, x^n)$; $i = 1, \ldots, n$; let $\tilde{f} = \tilde{f}(y^1, \ldots, y^n)$. Then

$$(F_*L)\,\tilde{f} = \sum_{r,s=1}^{n} \sum_{i,j=1}^{n} a_{ij} \frac{\partial F^r}{\partial x^i} \frac{\partial F^s}{\partial x^j} \frac{\partial^2 \tilde{f}}{\partial y^r \, \partial y^s} + \sum_{r=1}^{n} \sum_{i,j=1}^{n} \left(a_{ij} \frac{\partial^2 F^r}{\partial x^i \, \partial x^j} + a_i \frac{\partial F^r}{\partial x^i} \right) \frac{\partial \tilde{f}}{\partial y^r},$$

i.e., F_*L takes the form

$$F_*L = \sum_{r,s=1}^{n} b_{rs} \frac{\partial^2}{\partial y^r \, \partial y^s} + \sum_{r=1}^{n} b_r \frac{\partial}{\partial y^r} \tag{2.9}$$

with

$$b_{rs} = \sum_{i,j=1}^{n} a_{ij} \frac{\partial F^r}{\partial x^i} \frac{\partial F^s}{\partial x^j}, \qquad b_r = LF^r.$$

Thus F_*L is a second order operator on V. If $(\mathrm{d}F)_m: T_m(U) \to T_{F(m)}(V)$ is an isomorphism for each $m \in U$, we have $\det \|\partial F^i/\partial x^j\| \neq 0$, and *the ellipticity of L on U implies the ellipticity of F_*L on V.*

Now, we are in the position to present the following

Definition 2.1. *Let M be a manifold. A* second order (linear) operator *on M is a mapping L of the space of real-valued functions on M into itself with the following properties:* (i) *$\mu: U \to \mathbf{R}^n$ being a map of M and $f: M \to \mathbf{R}$ an arbitrary function, $Lf|_U$ is given by $L_{(\mu,U)}(f \circ \mu^{-1})$ with*

$$L_{(\mu,U)} = \sum_{i,j=1}^{n} a_{ij}(x) \frac{\partial^2}{\partial x^i \, \partial x^j} + \sum_{i=1}^{n} a_i(x) \frac{\partial}{\partial x^i}; \tag{2.10}$$

(ii) *$\mu: U \to \mathbf{R}^n$ and $\nu: V \to \mathbf{R}^n$ being two maps of M with $U \cap V \neq 0$, we have*

$$(\nu \circ \mu^{-1})_* \, L_{(\mu,U\cap V)} = L_{(\nu,U\cap V)}. \tag{2.11}$$

The operator L is called elliptic *on M if $L_{(\mu,U)}$ is elliptic for each map $\mu: U \to \mathbf{R}^n$ of M.*

In our applications to concrete problems of differential geometry, we shall need a very weak form of E. Hopf's maximum principle, namely

Theorem 2.2. *Let M be a manifold, ∂M its boundary, L an elliptic second order linear operator on M, $f: M \to \mathbf{R}$ a function. Suppose:* (i) *$f \geqq 0$ and $Lf \geqq 0$ on M;* (ii) *$f = 0$ on ∂M. Then $f = 0$ on M.*

The proof follows easily.

3. Pseudoanalytic functions

Let $G \subset \mathbf{R}^2$ be a bounded domain, ∂G its boundary. On G, be given functions $a_{ij}(x, y)$, $b_{ij}(x, y)$, $c_{ij}(x, y)$; $i, j = 1, 2$. On G, consider the system

$$\begin{aligned} a_{11} \frac{\partial f}{\partial x} + a_{12} \frac{\partial f}{\partial y} + b_{11} \frac{\partial g}{\partial x} + b_{12} \frac{\partial g}{\partial y} &= c_{11} f + c_{12} g, \\ a_{21} \frac{\partial f}{\partial x} + a_{22} \frac{\partial f}{\partial y} + b_{21} \frac{\partial g}{\partial x} + b_{22} \frac{\partial g}{\partial y} &= c_{21} f + c_{22} g \end{aligned} \tag{3.1}$$

for the functions $f, g: G \to \mathbf{R}$. This system is called *elliptic* if the form

$$\begin{aligned} \Phi = {} & (a_{12}b_{22} - a_{22}b_{12}) \mu^2 + (a_{11}b_{21} - a_{21}b_{11}) \nu^2 \\ & - (a_{11}b_{22} - a_{21}b_{12} + a_{12}b_{21} - a_{22}b_{11}) \mu\nu \end{aligned} \tag{3.2}$$

is definite. The first assertion of the theory is given by

Theorem 3.1. *On G, be given an elliptic system* (3.1). *If f, g are its solutions satisfying $f = g = 0$ on ∂G, then $f = g = 0$ in G.*

This is a restriction of the more general result which claims that the solutions f, g of an elliptic system (3.1) are either trivial or the set $\{(x, y) \in \overline{G}; f(x, y) = p(x, y) = 0\}$

consists of isolated points; here, there are no conditions on the boundary behavior of f, g. This theorem goes back to T. CARLEMAN; for its proof (and the proofs of the following theorems as well) under very weak regularity conditions see I. N. VEKUA [4]. In what follows, Theorem 3.1 will be quite sufficient.

Let $F: G \to F(G) \subset \mathbf{R}^2$ be a diffeomorphism given by

$$u = u(x, y), \quad v = v(x, y). \tag{3.3}$$

It is easy to see that the functions $f\big(u(x, y), v(x, y)\big)$, $g\big(u(x, y), v(x, y)\big)$ are solutions of (3.1) if and only if the functions $f(u, v)$, $g(u, v)$ are solutions of the system

$$\begin{aligned}
&\left(a_{11}\frac{\partial u}{\partial x} + a_{12}\frac{\partial u}{\partial y}\right)\frac{\partial f}{\partial u} + \left(a_{11}\frac{\partial v}{\partial x} + a_{12}\frac{\partial v}{\partial y}\right)\frac{\partial f}{\partial v} \\
&+ \left(b_{11}\frac{\partial u}{\partial x} + b_{12}\frac{\partial u}{\partial y}\right)\frac{\partial g}{\partial u} + \left(b_{11}\frac{\partial v}{\partial x} + b_{12}\frac{\partial v}{\partial y}\right)\frac{\partial g}{\partial v} = c_{11}f + c_{12}g, \\
&\left(a_{21}\frac{\partial u}{\partial x} + a_{22}\frac{\partial u}{\partial y}\right)\frac{\partial f}{\partial u} + \left(a_{21}\frac{\partial v}{\partial x} + a_{22}\frac{\partial v}{\partial y}\right)\frac{\partial f}{\partial v} \\
&+ \left(b_{21}\frac{\partial u}{\partial x} + b_{22}\frac{\partial u}{\partial y}\right)\frac{\partial g}{\partial u} + \left(b_{21}\frac{\partial v}{\partial x} + b_{22}\frac{\partial v}{\partial y}\right)\frac{\partial g}{\partial v} = c_{21}f + c_{22}g.
\end{aligned} \tag{3.4}$$

Let us write (3.2) in the form $\Phi = A\mu^2 - B\mu\nu + C\nu^2$. The form Φ^* associated to (3.4) is then given by

$$\begin{aligned}
&\Phi^* = A^*\mu^{*2} - B^*\mu^*\nu^* + C^*\nu^{*2}; \\
&A^* = C\left(\frac{\partial v}{\partial x}\right)^2 + B\frac{\partial v}{\partial x}\frac{\partial v}{\partial y} + A\left(\frac{\partial v}{\partial y}\right)^2, \\
&C^* = C\left(\frac{\partial u}{\partial x}\right)^2 + B\frac{\partial u}{\partial x}\frac{\partial u}{\partial y} + A\left(\frac{\partial u}{\partial y}\right)^2, \\
&B^* = 2C\frac{\partial u}{\partial x}\frac{\partial v}{\partial x} + B\left(\frac{\partial u}{\partial x}\frac{\partial v}{\partial y} + \frac{\partial u}{\partial y}\frac{\partial v}{\partial x}\right) + 2A\frac{\partial u}{\partial y}\frac{\partial v}{\partial y}.
\end{aligned} \tag{3.5}$$

From this,

$$4A^*C^* - B^{*2} = (4AC - B^2)\left(\frac{\partial u}{\partial x}\frac{\partial v}{\partial y} - \frac{\partial u}{\partial y}\frac{\partial v}{\partial x}\right)^2. \tag{3.6}$$

Thus: *The system* (3.1) *being elliptic, the system* (3.4) *is elliptic as well.* This enables us to define, for a general two-dimensional manifold M^2, elliptic systems taking, in each map $\mu: U \to \mathbf{R}^2$ of M^2, the form (3.1). I do not enter into the details which are easy to formulate. The following theorem gives us the information about the number of global solutions of such a system.

Theorem 3.2. *Let S^2 be a two-dimensional sphere with an elliptic system of the type described above. Let $m_0 \in S^2$ and $f_0, g_0 \in \mathbf{R}$. Then there are exactly two functions $f, g: S^2 \to \mathbf{R}$ such that the couple (f, g) is a solution of our system and $f(m_0) = f_0$, $g(m_0) = g_0$.*

For compact two-dimensional manifolds other than the sphere, this theorem is, generally, not valid.

Finally, let us consider special systems

$$(3.7) \qquad \frac{\partial f}{\partial x} - \frac{\partial g}{\partial y} = c_{11}f + c_{12}g, \qquad \frac{\partial f}{\partial y} + \frac{\partial g}{\partial x} = c_{21}f + c_{22}g$$

on a bounded domain $G \subset \mathbf{R}^2$. Because of $\Phi = \mu^2 + \nu^2$, each such system is elliptic. On $\mathbf{R}^2$, introduce the complex coordinate

$$(3.8) \qquad z = x + iy,$$

and let

$$(3.9) \qquad \frac{\partial}{\partial \bar{z}} = \frac{1}{2}\left(\frac{\partial}{\partial u} + \mathrm{i}\frac{\partial}{\partial v}\right)$$

be the well known operator. For

$$(3.10) \qquad w = f + \mathrm{i}g,$$

(3.7) may be written as

$$(3.11) \qquad \frac{\partial w}{\partial \bar{z}} + Aw + B\overline{w} = 0;$$

$$A = -\frac{1}{4}(c_{11} + c_{22}) + \frac{1}{4}\mathrm{i}(c_{12} - c_{21}),$$

$$B = \frac{1}{4}(c_{22} - c_{11}) - \frac{1}{4}\mathrm{i}(c_{12} + c_{21}).$$

One of the main problems of the theory of systems of the form (3.7) may be formulated as the so-called

Problem $\mathbf{A}^0$. *We have to investigate the existence of solutions $w: \overline{G} \to \mathbf{C}$ satisfying* (3.11) *in G and the boundary condition*

$$(3.12) \qquad \operatorname{Re}\left[\overline{\lambda(z)}\, w(z)\right] = 0 \quad on\ \partial G.$$

Here, $A, B: G \to \mathbf{C}$ and $\lambda: \partial G \to \mathbf{C}$ are given, and λ satisfies

$$(3.13) \qquad |\lambda(z)| = 1 \quad on\ \partial G.$$

Suppose that G is an $(m + 1)$-connected domain, and let its boundary ∂G consist of $m + 1$ components $\Gamma_0, \Gamma_1, \ldots, \Gamma_m$. Let

$$(3.14) \qquad n = n_0 + \cdots + n_m; \qquad n_j = \frac{1}{2\pi} \Delta_{\Gamma_j} \arg \lambda(t) \equiv \operatorname{ind}_{\Gamma_j} \lambda;$$

be the *index* of the problem $\mathbf{A}^0$. Recall, how to calculate n_j: take a point $m_0 \in \Gamma_j$, and run once through Γ_j (i.e., returning to m_0) in such a direction that G remains on the left-hand side; n_j is then equal to the total number of positive turns of the unit vector λ.

The main results are now forn ulated.

Theorem 3.3. *Let $w(z)$ be a non-trivial (in the sense $w(z) \not\equiv 0$) solution of $\mathbf{A}^0$. Then there is a polynomial $P(z)$ with roots in $\overline{G}$ and a function $\tilde{w}: \overline{G} \to \mathbf{C}$ with $w(z) \neq 0$ for each $z \in \overline{G}$ such that*

$$(3.15) \qquad w(z) = P(z)\, \tilde{w}(z).$$

Because of this, we may speak about the multiplicity of the zeros of a non-trivial solution of $\mathbf{A}^0$; by this we mean the multiplicity of the zeros of $P(z)$. Thus we see that the zeros of (3.7) behave like the zeros of a polynomial; in the more general case (3.1), we have a similar representation (3.15), $P(z)$ being now a holomorphic function. •

Theorem 3.4. (i) *Let $n < 0$. Then $\mathbf{A}^0$ has no non-trivial solutions.* (ii) *Let $n = 0$. If $\mathbf{A}^0$ has a non-trivial solution, then all solutions of $\mathbf{A}^0$ are of the form $w(z) = cw_0(z)$ with $c \in \mathbf{R}$ and $w_0(z)$ a solution of $\mathbf{A}^0$ such that $w_0(z) \neq 0$ for each $z \in \overline{G}$.* (iii) *Let $n > 0$. If $\mathbf{A}^0$ has a non-trivial solution $w(z)$ and N_G and $N_{\partial G}$ is the number of its zeros on G and ∂G resp., we have $N_{\partial G} + 2N_G = 2n$. On each Γ_j, the number of the zeros of $w(z)$ is even.* (We have to take into account the multiplicities of the zeros.)

Theorem 3.5. *Let l denote the number of linearly independent* (*over* $\mathbf{R}$) *solutions of $\mathbf{A}^0$.* (i) *Let $n > m - 1$. Then $l = 2n + 1 - m$.* (ii) *Let $0 \leqq n < \frac{1}{2} m$. Then $0 \leqq l \leqq n + 1$.* (iii) *Let $\frac{1}{2} m \leqq n \leqq m - 1$. Then $2n + 1 - m \leqq l \leqq 3n + 2 - m$.*

In the inequalities for l in Theorem 3.5, all possibilities might be attained.

IV. Local differential geometry of surfaces in E^3

1. Structure equations

First of all, let us explain several general conventions used throughout the text.

We shall restrict ourselves to domains $\mathscr{D} \subset \mathbf{R}^n$ with the following property: there is a domain $\mathscr{D}' \subset \mathbf{R}^n$ such that $\overline{\mathscr{D}} \subset \mathscr{D}'$ and to each point $d \in \partial\mathscr{D} = \overline{\mathscr{D}} - \mathscr{D}$ there are local coordinates $(y^1, \ldots, y^n)$ in some neighbourhood $U_d \subset \mathbf{R}^n$ of d with $\partial\mathscr{D} \cap U_d = \{(y^1, \ldots, y^n); y^n = 0\}$, $\mathscr{D} \cap U_d = \{(y^1, \ldots, y^n); y^n > 0\}$.

By a manifold with a boundary $M: \mathscr{D} \to E^n$ we shall mean a mapping $M: \mathscr{D}' \to E^n$ such that $\mathscr{D}'$ is some domain containing $\mathscr{D}$ in the above sense and $(\mathrm{d}M)_{d'}$ has the maximal rank for each $d' \in \mathscr{D}'$. For abuse of the language, we shall often identify $M(\mathscr{D})$ with M and $M(\partial\mathscr{D})$ with ∂M resp., the reader is invited to make all the considerations quite precise. Thus, for example, the sentence "If M has the property $\mathscr{P}_1$ and ∂M has the property $\mathscr{P}_2$, then M has the property $\mathscr{P}_3$" should read "If $M(\mathscr{D}')$ has the property $\mathscr{P}_1$ for some $\mathscr{D}' \supset \mathscr{D}$ as above and $M(\partial\mathscr{D})$ has the property $\mathscr{P}_2$, then $M(\overline{\mathscr{D}})$ has the property $\mathscr{P}_3$". I am quite sure that there is no need for a more carefull treatment.

Let $\mathscr{D} \subset \mathbf{R}^m$ be a domain and $m: \mathscr{D} \to E^n$ a mapping of $\mathscr{D}$ into the Euclidean space E^n. To each point $x \in \mathscr{D}$ associate an orthonormal frame $\{m(x), v_1(x), \ldots, v_n(x)\}$ of E^n. Then we may write

$$(1.1) \qquad \mathrm{d}m(x) = \sum_{i=1}^{n} \omega^i v_i(x), \qquad \mathrm{d}v_i(x) + \sum_{j=1}^{n} \omega_i{}^j v_j(x),$$

ω^i, $\omega_i{}^j$ being 1-forms on $\mathscr{D}$. The differential consequences of (1.1) are (d is the exterior differential operator!)

$$0 = \mathrm{d}^2 m = \sum_{i=1}^{n} (\mathrm{d}v_i \wedge \omega^i + v_i\, \mathrm{d}\omega^i) = \sum_{i,j=1}^{n} v_j \omega_i{}^j \wedge \omega^i + \sum_{i=1}^{n} v_i\, \mathrm{d}\omega^i$$
$$= \sum_{i=1}^{n} v_i \left(\sum_{j=1}^{n} \omega_j{}^i \wedge \omega^j + \mathrm{d}\omega^i \right),$$
$$0 = \mathrm{d}^2 v_i = \sum_{j=1}^{n} (\mathrm{d}v_j \wedge \omega_i{}^j + v_j\, \mathrm{d}\omega_i{}^j) = \sum_{j,k=1}^{n} v_k \omega_j{}^k \wedge \omega_i{}^j + \sum_{j=1}^{n} v_j\, \mathrm{d}\omega_i{}^j$$
$$= \sum_{j=1}^{n} v_j \left(\sum_{k=1}^{n} \omega_k{}^j \wedge \omega_i{}^k + \mathrm{d}\omega_i{}^j \right),$$

because of the linear independence of the vectors v_i,

$$(1.2) \qquad \mathrm{d}\omega^i = \sum_{j=1}^{n} \omega^j \wedge \omega_j{}^i, \quad \mathrm{d}\omega_i{}^j = \sum_{k=1}^{n} \omega_i{}^k \wedge \omega_k{}^j \qquad (i, j = 1, \ldots, n).$$

These are the so-called *structure equations*.

Let $\langle\,,\rangle$ denote the scalar product in E^n. We have supposed the frames to be orthonormal, i.e.,

$$(1.3)\qquad \langle v_i, v_j\rangle = \delta_{ij} = \begin{cases} 1 & \text{for} \quad i = j, \\ 0 & \text{for} \quad i \neq j. \end{cases}$$

Thus

$$\left\langle \sum_{k=1}^{n} \omega_i^{\ k} v_k, v_j \right\rangle + \left\langle v_i, \sum_{k=1}^{n} \omega_j^{\ k} v_k \right\rangle = \sum_{k=1}^{n} (\omega_i^{\ k}\delta_{kj} + \omega_j^{\ k}\delta_{ik}) = 0,$$

i.e.,

$$(1.4)\qquad \omega_i^{\ j} + \omega_j^{\ i} = 0 \qquad (i, j = 1, \ldots, n).$$

Let $\{m(x), {}^*v_1(x), \ldots, {}^*v_n(x)\}$ be another field of orthonormal frames, and let us suppose

$$(1.5)\qquad {}^*v_i = \sum_{j=1}^{n} A_i^{\ j} v_j;$$

of course, $(A_i^{\ j})$ is an orthogonal matrix, i.e.,

$$(1.6)\qquad \delta_{ij} = \sum_{k,l=1}^{n} \delta_{kl} A_i^{\ k} A_j^{\ l} \qquad (i, j = 1, \ldots, n).$$

Let $(\tilde{A}_i^{\ j})$ be the matrix inverse to $(A_i^{\ j})$, i.e.,

$$(1.7)\qquad \sum_{k=1}^{n} A_i^{\ k} \tilde{A}_k^{\ j} = \sum_{k=1}^{n} A_k^{\ j} \tilde{A}_i^{\ k} = \delta_i^{\ j} = \begin{cases} 1 & \text{for} \quad i = j, \\ 0 & \text{for} \quad i \neq j. \end{cases}$$

In the new frames, we have

$$(1.8)\qquad \mathrm{d}m = \sum_{i=1}^{n} {}^*\omega^i\, {}^*v_i, \quad \mathrm{d}{}^*v_i = \sum_{j=1}^{n} {}^*\omega_i^{\ j}\, {}^*v_j \quad (i = 1, \ldots, n).$$

Substituting (1.5) into (1.8), we get

$$\sum_{j=1}^{n} \omega^j v_j = \sum_{j=1}^{n} \left(\sum_{i=1}^{n} {}^*\omega^i A_i^{\ j} \right) v_j,$$

$$\sum_{k=1}^{n} \left(\mathrm{d}A_i^{\ k} + \sum_{j=1}^{n} A_i^{\ j} \omega_j^{\ k} \right) v_k = \sum_{k=1}^{n} \left(\sum_{j=1}^{n} A_j^{\ k}\, {}^*\omega_i^{\ j} \right) v_k,$$

i.e.,

$$\omega^j = \sum_{i=1}^{n} A_i^{\ j}\, {}^*\omega^i, \quad \mathrm{d}A_i^{\ k} + \sum_{j=1}^{n} A_i^{\ j} \omega_j^{\ k} = \sum_{j=1}^{n} A_j^{\ k}\, {}^*\omega_i^{\ j}$$

and

$$(1.9)\qquad {}^*\omega^i = \sum_{j=1}^{n} \tilde{A}_j^{\ i} \omega^j, \quad {}^*\omega_i^{\ j} = \sum_{k=1}^{n} \tilde{A}_k^{\ j} \left(\mathrm{d}A_i^{\ k} + \sum_{l=1}^{n} A_i^{\ l} \omega_l^{\ k} \right) \quad (i, j = 1, \ldots, n).$$

2. Surfaces in E^3

Let $M : \mathscr{D} \to E^3$ be a surface in E^3. To each point $d \in \mathscr{D}$, let us associate a frame $\{m, v_1, v_2, v_3\}$ of E^3 such that $m = M(d)$ and v_1, v_2 are tangent to M at m (these are the so-called *tangent frames*). Then

$$(2.1)\qquad \begin{aligned} dm &= \omega^1 v_1 + \omega^2 v_2, \\ dv_1 &= \qquad\quad \omega_1{}^2 v_2 + \omega_1{}^3 v_3, \\ dv_2 &= -\omega_1{}^2 v_1 \qquad\quad + \omega_2{}^3 v_3, \\ dv_3 &= -\omega_1{}^3 v_1 - \omega_2{}^3 v_2 \end{aligned}$$

because of (1.4). The structure equations (1.2) reduce to

$$(2.2)\qquad \omega^1 \wedge \omega_1{}^3 + \omega^2 \wedge \omega_2{}^3 = 0,$$

$$(2.3)\qquad d\omega^1 = -\omega^2 \wedge \omega_1{}^2, \quad d\omega^2 = \omega^1 \wedge \omega_1{}^2,$$

$$(2.4)\qquad d\omega_1{}^2 = -\omega_1{}^3 \wedge \omega_2{}^3, \quad d\omega_1{}^3 = \omega_1{}^2 \wedge \omega_2{}^3, \quad d\omega_2{}^3 = -\omega_1{}^2 \wedge \omega_1{}^3.$$

Of course, the 1-forms ω^1, ω^2 are linearly independent, i.e.,

$$(2.5)\qquad do := \omega^1 \wedge \omega^2 \neq 0;$$

do is the so-called *volume element* of M.

Let $\{m, {}^*v_1, {}^*v_2, {}^*v_3\}$ be another field of tangent frames; let

$$(2.6)\qquad \begin{aligned} {}^*v_1 &= \varepsilon_1(\cos\varphi \cdot v_1 - \sin\varphi \cdot v_2), \\ {}^*v_2 &= \sin\varphi \cdot v_1 + \cos\varphi \cdot v_2, \\ {}^*v_3 &= \varepsilon_2 v_3; \quad \varepsilon_1{}^2 = \varepsilon_2{}^2 = 1; \end{aligned}$$

the equations analogous to (2.1) being

$$(2.7)\qquad \begin{aligned} dm &= {}^*\omega^1 \cdot {}^*v_1 + {}^*\omega^2 \cdot {}^*v_2, \\ d{}^*v_1 &= {}^*\omega_1{}^2 \cdot {}^*v_2 + {}^*\omega_1{}^3 \cdot {}^*v_3, \\ d{}^*v_2 &= -{}^*\omega_1{}^2 \cdot {}^*v_1 + {}^*\omega_2{}^3 \cdot {}^*v_3, \\ d{}^*v_3 &= -{}^*\omega_1{}^3 \cdot {}^*v_1 - {}^*\omega_2{}^3 \cdot {}^*v_2. \end{aligned}$$

Inserting (2.6) into (2.7) and taking regard of (2.1), we get

$$(2.8)\qquad \omega^1 = \varepsilon_1 \cos\varphi \cdot {}^*\omega^1 + \sin\varphi \cdot {}^*\omega^2, \quad \omega^2 = -\varepsilon_1 \sin\varphi \cdot {}^*\omega^1 + \cos\varphi \cdot {}^*\omega^2,$$

$$(2.9)\qquad {}^*\omega^1 = \varepsilon_1(\cos\varphi \cdot \omega^1 - \sin\varphi \cdot \omega^2), \quad {}^*\omega^2 = \sin\varphi \cdot \omega^1 + \cos\varphi \cdot \omega^2,$$

$$(2.10)\qquad {}^*\omega_1{}^2 = \varepsilon_1(\omega_1{}^2 - d\varphi),$$

$$(2.11)\qquad {}^*\omega_1{}^3 = \varepsilon_1\varepsilon_2(\cos\varphi \cdot \omega_1{}^3 - \sin\varphi \cdot \omega_2{}^3), \quad {}^*\omega_2{}^3 = \varepsilon_2(\sin\varphi \cdot \omega_1{}^3 + \cos\varphi \cdot \omega_2{}^3).$$

Introduce the following quadratic differential forms:

$$(2.12)\qquad I = \langle dm, dm\rangle = (\omega^1)^2 + (\omega^2)^2,$$

$$(2.13)\qquad II = -\langle dm, dv_3\rangle = \omega^1\omega_1{}^3 + \omega^2\omega_2{}^3,$$

$$(2.14)\qquad III = \langle \mathrm{d}v_3, \mathrm{d}v_3 \rangle = (\omega_1{}^3)^2 + (\omega_2{}^3)^2$$

with $\langle .,. \rangle$ again the scalar product in E^3.

Definition 2.1. *The forms I, II, III are called the* first, second *and* third fundamental forms *of our surface resp.*

From (2.8)—(2.11), we get the following

Lemma 2.1. *We have*

$$(2.15)\qquad {}^*I = I,\ {}^*II = \varepsilon_2 II,\quad {}^*III = III;$$

$$(2.16)\qquad \mathrm{d}^*o = \varepsilon_1\, \mathrm{d}o.$$

Applying the Cartan's lemma I.3.13 to (2.2), we obtain the existence of functions a, b, c on M such that

$$(2.17)\qquad \omega_1{}^3 = a\omega^1 + b\omega^2,\quad \omega_2{}^3 = b\omega^1 + c\omega^2.$$

From (2.9)—(2.11),

$$(2.18)\qquad \begin{aligned} {}^*a &= \varepsilon_2(\cos^2\varphi\cdot a - \sin 2\varphi\cdot b + \sin^2\varphi\cdot c),\\ {}^*b &= \frac{1}{2}\,\varepsilon_1\varepsilon_2(\sin 2\varphi\cdot a + 2\cos 2\varphi\cdot b - \sin 2\varphi\cdot c),\\ {}^*c &= \varepsilon_2(\sin^2\varphi\cdot a + \sin 2\varphi\cdot b + \cos^2\varphi\cdot c). \end{aligned}$$

Introducing the functions

$$(2.19)\qquad H = \frac{1}{2}\,(a + c),\quad K = ac - b^2,$$

we obtain the following

Lemma 2.2. *We have*

$$(2.20)\qquad {}^*H = \varepsilon_2 H,\quad {}^*K = K.$$

Definition 2.2. *The functions H, K* (2.19) *are the so-called* mean *and* Gauss curvatures *of our surface resp.*

Let us remark that H is not an invariant in the proper sense (we should introduce the invariant H^2), but I do follow the historical custom.

The exterior differentiation of (2.17) implies

$$(2.21)\qquad \begin{aligned} &(\mathrm{d}a - 2b\omega_1{}^2)\wedge\omega^1 + \{\mathrm{d}b + (a - c)\,\omega_1{}^2\}\wedge\omega^2 = 0,\\ &\{\mathrm{d}b + (a - c)\,\omega_1{}^2\}\wedge\omega^1 + (\mathrm{d}c + 2b\omega_1{}^2)\wedge\omega^2 = 0, \end{aligned}$$

and we get the existence of functions $\alpha, \beta, \gamma, \delta$ such that

$$(2.22)\qquad \begin{aligned} &\mathrm{d}a - 2b\omega_1{}^2 = \alpha\omega^1 + \beta\omega^2,\\ &\mathrm{d}b + (a - c)\,\omega_1{}^2 = \beta\omega^1 + \gamma\omega^2,\\ &\mathrm{d}c + 2b\omega_1{}^2 = \gamma\omega^1 + \delta\omega^2. \end{aligned}$$

A further differentiation of (2.22) implies

$$(2.23) \quad \begin{aligned} &(\mathrm{d}\alpha - 3\beta\omega_1{}^2) \wedge \omega^1 + \{\mathrm{d}\beta + (\alpha - 2\gamma)\,\omega_1{}^2\} \wedge \omega^2 = 2bK\omega^1 \wedge \omega^2, \\ &\{\mathrm{d}\beta + (\alpha - 2\gamma)\,\omega_1{}^2\} \wedge \omega^1 + \{\mathrm{d}\gamma + (2\beta - \delta)\,\omega_1{}^2\} \wedge \omega^2 = (c - a)\,K\omega^1 \wedge \omega^2, \\ &\{\mathrm{d}\gamma + (2\beta - \delta)\,\omega_1{}^2\} \wedge \omega^1 + (\mathrm{d}\delta + 3\gamma\omega_1{}^2) \wedge \omega^2 = -2bK\omega^1 \wedge \omega^2 \end{aligned}$$

and the existence of functions $A, \ldots, E$ such that

$$(2.24) \quad \begin{aligned} &\mathrm{d}\alpha - 3\beta\omega_1{}^2 = A\omega^1 + (B - bK)\,\omega^2, \\ &\mathrm{d}\beta + (\alpha - 2\gamma)\,\omega_1{}^2 = (B + bK)\,\omega^1 + (C + aK)\,\omega^2, \\ &\mathrm{d}\gamma + (2\beta - \delta)\,\omega_1{}^2 = (C + cK)\,\omega^1 + (D + bK)\,\omega^2, \\ &\mathrm{d}\delta + 3\gamma\omega_1{}^2 = (D - bK)\,\omega^1 + E\omega^2. \end{aligned}$$

The relations between $\alpha, \ldots, E$ and $^*\alpha, \ldots, {}^*E$ will be established later on. Of course, they are many ways how to solve (2.23); one of them is, e.g.,

$$(2.25) \quad \begin{aligned} &\mathrm{d}\alpha - 3\beta\omega_1{}^2 = A'\omega^1 + B'\omega^2, \\ &\mathrm{d}\beta + (\alpha - 2\gamma)\,\omega_1{}^2 = (B' + 2bK)\,\omega^1 + (C' + aK)\,\omega^2, \\ &\mathrm{d}\gamma + (2\beta - \delta)\,\omega_1{}^2 = (C' + cK)\,\omega^1 + (D' + 2bK)\,\omega^2, \\ &\mathrm{d}\delta + 3\gamma\omega_1{}^2 = D'\omega^1 + E'\omega^2; \end{aligned}$$

nevertheless, I prefer (2.24).

Let us remark that

$$(2.26) \quad H^2 - K = \frac{1}{4}(a - c)^2 + b^2 \geqq 0.$$

Definition 2.3. *The point $m \in M$ is called* umbilical *if $H^2 - K = 0$ at m.*

Proposition 2.1. *Let M consist of umbilical points. Then M is a part of a sphere or a part of a plane resp.*

Proof. Each point of M being umbilical, we have

$$(2.27) \quad a = c, \qquad b = 0$$

on M. Substituting this into (2.22), we get

$$(2.28) \quad \alpha = \beta = \gamma = \delta = 0$$

and $\mathrm{d}a = 0$, i.e., $a = \text{const}$ on M. In the case $a = 0$, we have $\mathrm{d}v_3 = 0$, i.e., $v_3 = \text{const}$, and M is a part of a plane. Let $a \neq 0$. Then

$$(2.29) \quad \mathrm{d}(m + a^{-1}v_3) = 0,$$

and it is easy to see that M is a part of a sphere centered at the point $m + a^{-1}v_3$ and having the radius a^{-1}. QED.

Definition 2.4. *Let v be a non-zero tangent vector of M at m. Its* normal curvature *is defined as*

$$(2.30) \qquad k_n(v) = \frac{II(v)}{I(v)}.$$

The vectors v with $k_n(v)$ extremal are called the principal vectors, *the corresponding normal curvatures are called the* principal curvatures. *A curve on M is called* principal *if each its tangent vector is principal.*

Because of (2.15_2), $k_n(v)$ is defined up to the sign.

Proposition 2.2. *There are exactly two possibilities:* (1) *each tangent vector v at m is principal and m is umbilical,* (2) *there are exactly two mutually orthogonal unit principial vectors at m and each other principial vector at m is a multiple of one of them.*

Proof. Let us choose a field of frames as above, and let $v = xv_1 + yv_2$. Then — see (2.13) and (2.17) —

$$(2.31) \qquad k_n(v) = \frac{ax^2 + 2bxy + cy^2}{x^2 + y^2},$$

The normal curvature being extremal, we have

$$(2.32) \qquad k_n(v) \cdot x = ax + by, \qquad k_n(v) \cdot y = bx + cy.$$

The elimination of $k_n(v)$ leads to

$$(2.33) \qquad bx^2 + (c - a)\,xy - by^2 = 0,$$

the elimination of x, y to

$$(2.34) \qquad k_n(v)^2 - 2Hk_n(v) + K = 0.$$

Our results follows from (2.33). QED.

Let us notice the following auxiliary result. In each case, we may choose the frames (at least in a suitable neighbourhood of m) in such a way that v_1 and v_2 are principial vectors at m. Then $b(m) = 0$ and $k_n(v_1) = a$, $k_n(v_2) = c$ are the principial curvatures at m. In general, there is no such field of frames satisfying $b = 0$ at each point $m \in M$: take, as an example, for M a small neighbourhood of the vertex of an ellipsoid of revolution.

Theorem 2.1. *Let $M, \tilde{M} : \mathscr{D} \to E^3$ be two isometric surfaces, i.e., let them satisfy*

$$(2.35) \qquad I = \tilde{I}.$$

Then

$$(2.36) \qquad K = \tilde{K}.$$

Proof. The frames $\{m, v_1, v_2, v_3\}$, $\{\tilde{m}_1, \tilde{v}_1, \tilde{v}_2, \tilde{v}_3\}$ of M or $\tilde{M}$ resp. be chosen in such a way that $\tilde{v}_i = \{\mathrm{d}(\tilde{M} \circ M^{-1})\}\, v_i$ for $i = 1, 2$. Then (2.35) implies

$$(2.37) \qquad \mathrm{d}\tilde{m} = \omega^1\tilde{v}_1 + \omega^2\tilde{v}_1, \qquad \mathrm{d}\tilde{v}_1 = \tilde{\omega}_1{}^2\tilde{v}_2 + \tilde{\omega}_1{}^3\tilde{v}_3,$$
$$\mathrm{d}\tilde{v}_2 = -\tilde{\omega}_1{}^2\tilde{v}_1 + \tilde{\omega}_2{}^3\tilde{v}_3, \qquad \mathrm{d}\tilde{v}_3 = -\tilde{\omega}_1{}^3\tilde{v}_1 - \tilde{\omega}_2{}^3\tilde{v}_3;$$

let us write

$$(2.38)\qquad \tilde{\omega}_1{}^3 = \tilde{a}\omega^1 + \tilde{b}\omega^2, \qquad \tilde{\omega}_2{}^3 = \tilde{b}\omega^1 + \tilde{c}\omega^2.$$

Then

$$d\omega^1 = -\omega^2 \wedge \omega_1{}^2 = -\omega^2 \wedge \tilde{\omega}_1{}^2, \qquad d\omega^2 = \omega^1 \wedge \omega_1{}^2 = \omega^1 \wedge \tilde{\omega}_1{}^2,$$

i.e.,

$$(\omega_1{}^2 - \tilde{\omega}_1{}^2) \wedge \omega^1 = (\omega_1{}^2 - \tilde{\omega}_1{}^2) \wedge \omega^2 = 0,$$

i.e.,

$$(2.39)\qquad \omega_1{}^2 = \tilde{\omega}_1{}^2.$$

Thus $d\omega_1{}^2 = -\omega_1{}^3 \wedge \omega_2{}^3 = -\tilde{\omega}_1{}^3 \wedge \tilde{\omega}_2{}^3$, i.e., $(ac - b^2)\,\omega^1 \wedge \omega^2 = (\tilde{a}\tilde{c} - \tilde{b}^2)\,\omega^1 \wedge \omega^2$, i.e., (2.36). QED.

Theorem 2.2. *Let $M, \tilde{M} : \mathscr{D} \to E^3$ be two isometric surfaces satisfying*

$$(2.40)\qquad II = \widetilde{II}.$$

Then there is a motion $\mu: E^3 \to E^3$ such that $\mu \circ M = \tilde{M}$.

Proof. We have (2.39) and $\tilde{\omega}_1{}^3 = \omega_1{}^3$, $\tilde{\omega}_2{}^3 = \omega_2{}^3$ because of (2.40). QED.

3. Forms on a surface

Let $M: \mathscr{D} \to E^3$ be a surface, let $\{m, v_1, v_2, v_3\}$, $\{m, {}^*v_1, {}^*v_2, {}^*v_3\}$ be two associated fields of frames. Let ω be a 1-form on M; it may be expressed as

$$(3.1)\qquad \omega = R_1\omega^1 + R_2\omega^2 = {}^*R_1{}^*\omega^1 + {}^*R_2{}^*\omega^2.$$

Because of (2.8), (2.9),

$$(3.2)\qquad {}^*R_1 = \varepsilon_1(\cos\varphi \cdot R_1 - \sin\varphi \cdot R_2), \qquad {}^*R_2 = \sin\varphi \cdot R_1 + \cos\varphi \cdot R_2.$$

Define the *covariant derivatives* R_{ij} of R_i with respect to the field $\{m, v_1, v_2, v_3\}$ by means of the formulas

$$(3.3)\qquad \begin{aligned} dR_1 - R_2\omega_1{}^2 &= R_{11}\omega^1 + R_{12}\omega^2, \\ dR_2 + R_1\omega_1{}^2 &= R_{21}\omega^1 + R_{22}\omega^2. \end{aligned}$$

By an easy calculation, we obtain

$$\begin{aligned} \varepsilon_1 \cos\varphi \cdot R_{11} - \varepsilon_1 \sin\varphi \cdot R_{21} &= \varepsilon_1 \cos\varphi \cdot {}^*R_{11} + \sin\varphi \cdot {}^*R_{12}, \\ \varepsilon_1 \cos\varphi \cdot R_{12} - \varepsilon_1 \sin\varphi \cdot R_{22} &= -\varepsilon_1 \sin\varphi \cdot {}^*R_{11} + \cos\varphi \cdot {}^*R_{12}, \\ \sin\varphi \cdot R_{11} + \cos\varphi \cdot R_{21} &= \varepsilon_1 \cos\varphi \cdot {}^*R_{21} + \sin\varphi \cdot {}^*R_{22}, \\ \sin\varphi \cdot R_{12} + \cos\varphi \cdot R_{22} &= -\varepsilon_1 \sin\varphi \cdot {}^*R_{21} + \cos\varphi \cdot {}^*R_{22}, \end{aligned}$$

i.e.,

$$(3.4)\qquad \begin{aligned} {}^*R_{11} &= \cos^2\varphi \cdot R_{11} - \sin\varphi\cos\varphi \cdot R_{12} - \sin\varphi\cos\varphi \cdot R_{21} + \sin^2\varphi \cdot R_{22}, \\ {}^*R_{12} &= \varepsilon_1(\sin\varphi\cos\varphi \cdot R_{11} + \cos^2\varphi \cdot R_{12} - \sin^2\varphi \cdot R_{21} - \sin\varphi\cos\varphi \cdot R_{22}), \\ {}^*R_{21} &= \varepsilon_1(\sin\varphi\cos\varphi \cdot R_{11} - \sin^2\varphi \cdot R_{12} + \cos^2\varphi \cdot R_{21} - \sin\varphi\cos\varphi \cdot R_{22}), \\ {}^*R_{22} &= \sin^2\varphi \cdot R_{11} + \sin\varphi\cos\varphi \cdot R_{12} + \sin\varphi\cos\varphi \cdot R_{21} + \cos^2\varphi \cdot R_{22}. \end{aligned}$$

From (3.2) and (3.4),

$$(3.5)\quad \begin{aligned} {}^*R_1{}^*R_{21} - {}^*R_2{}^*R_{11} &= \cos\varphi\cdot(R_1R_{21} - R_2R_{11}) - \sin\varphi\cdot(R_1R_{22} - R_2R_{12}),\\ {}^*R_1{}^*R_{22} - {}^*R_2{}^*R_{12} &= \varepsilon_1\sin\varphi\cdot(R_1R_{21} - R_2R_{11}) + \varepsilon_1\cos\varphi\cdot(R_1R_{22} - R_2R_{12}). \end{aligned}$$

Thus we get

Lemma 3.1. *To the* 1-*form* ω *on* M, *let us associate the functions*

$$(3.6)\quad G_\omega = R_1^2 + R_2^2, \qquad I_\omega = R_{11} + R_{22}, \qquad J_\omega = R_{11}R_{22} - R_{12}R_{21}$$

and the 1-*form*

$$(3.7)\quad \tilde\omega = (R_1R_{21} - R_2R_{11})\,\omega^1 + (R_1R_{22} - R_2R_{12})\,\omega^2.$$

Then

$$(3.8)\quad {}^*G_\omega = G_\omega, \qquad {}^*I_\omega = I_\omega, \qquad {}^*J_\omega = J_\omega,$$

$$(3.9)\quad {}^*\tilde\omega = \varepsilon_1\tilde\omega.$$

The further exterior differentiation of (3.3) yields

$$(3.10)\quad \begin{aligned} &\{\mathrm{d}R_{11} - (R_{12} + R_{21})\,\omega_1{}^2\}\wedge\omega^1 + \{\mathrm{d}R_{12} + (R_{11} - R_{22})\,\omega_1{}^2\}\wedge\omega^2\\ &= R_2K\omega^1\wedge\omega^2,\\ &\{\mathrm{d}R_{21} + (R_{11} - R_{22})\,\omega_1{}^2\}\wedge\omega^1 + \{\mathrm{d}R_{22} + (R_{12} + R_{21})\,\omega_1{}^2\}\wedge\omega^2\\ &= -R_1K\omega^1\wedge\omega^2, \end{aligned}$$

and we get the existence of functions $R_{ijk} = R_{ikj}$ such that

$$(3.11)\quad \begin{aligned} \mathrm{d}R_{11} - (R_{12} + R_{21})\,\omega_1{}^2 &= R_{111}\omega^1 + \left(R_{112} - \frac{1}{2}KR_2\right)\omega^2,\\ \mathrm{d}R_{12} + (R_{11} - R_{22})\,\omega_1{}^2 &= \left(R_{112} + \frac{1}{2}KR_2\right)\omega^1 + R_{122}\omega^2,\\ \mathrm{d}R_{21} + (R_{11} - R_{22})\,\omega_1{}^2 &= R_{211}\omega^1 + \left(R_{212} + \frac{1}{2}KR_1\right)\omega^2,\\ \mathrm{d}R_{22} + (R_{12} + R_{21})\,\omega_1{}^2 &= \left(R_{212} - \frac{1}{2}KR_1\right)\omega^1 + R_{222}\omega^2. \end{aligned}$$

By a direct calculation, we get

Lemma 3.2. *Let* ω *be a* 1-*form on* M, *then*

$$(3.12)\quad \mathrm{d}\tilde\omega = (2J_\omega - G_\omega K)\,\mathrm{d}o,$$

i.e., $\mathrm{d}\tilde\omega$ *does not depend on the functions* R_{ijk}.

On the set of 1-forms on M, let us introduce the well-known $*$-operator by means of the formula

$$(3.13)\quad *(R_1\omega^1 + R_2\omega^2) = -R_2\omega^1 + R_1\omega^2.$$

This definition is "almost" invariant, namely, we have

Lemma 3.3. *Let ω be a 1-form (3.1) on M. Then*

$$* (^*\omega) = \varepsilon_1 * \omega. \tag{3.14}$$

Proof. From (3.2) and (2.9),

$$\begin{aligned} * (^*R_1{}^*\omega^1 + {}^*R_2{}^*\omega^2) &= -{}^*R_2{}^*\omega^1 + {}^*R_1{}^*\omega^2 = \varepsilon_1(-R_2\omega^1 + R_1\omega^2) \\ &= \varepsilon_1 * (R_1\omega^1 + R_2\omega^2). \end{aligned}$$

QED.

Now, let us consider a function f on M. Then its first order covariant derivatives f_i with respect to a given field of frames are given by

$$\mathrm{d}f = f_1\omega^1 + f_2\omega^2. \tag{3.15}$$

The differential consequences of (3.15) being

$$(\mathrm{d}f_1 - f_2\omega_1{}^2) \wedge \omega^1 + (\mathrm{d}f_2 + f_1\omega_1{}^2) \wedge \omega^2 = 0, \tag{3.16}$$

we get the existence of second order covariant derivatives $f_{ij} = f_{ji}$ such that

$$\begin{aligned} \mathrm{d}f_1 - f_2\omega_1{}^2 &= f_{11}\omega^1 + f_{12}\omega^2, \\ \mathrm{d}f_2 + f_1\omega_1{}^2 &= f_{12}\omega^1 + f_{22}\omega^2, \end{aligned} \tag{3.17}$$

in other words, we may apply, on the form $\mathrm{d}f$, the procedure already used in handling a general 1-form (3.1). Of course, the equations (3.10) and (3.11) reduce to

$$\begin{aligned} (\mathrm{d}f_{11} - 2f_{12}\omega_1{}^2) \wedge \omega^1 + \{\mathrm{d}f_{12} + (f_{11} - f_{22})\, \omega_1{}^2\} \wedge \omega^2 &= f_2K\omega^1 \wedge \omega^2, \\ \{\mathrm{d}f_{12} + (f_{11} - f_{22})\, \omega_1{}^2\} \wedge \omega^1 + (\mathrm{d}f_{22} + 2f_{12}\omega_1{}^2) \wedge \omega^2 &= -f_1K\omega^1 \wedge \omega^2, \end{aligned} \tag{3.18}$$

$$\begin{aligned} \mathrm{d}f_{11} - 2f_{12}\omega_1{}^2 &= f_{111}\omega^1 + \left(f_{112} - \frac{1}{2} f_2K\right) \omega^2, \\ \mathrm{d}f_{12} + (f_{11} - f_{22})\, \omega_1{}^2 &= \left(f_{112} + \frac{1}{2} f_2K\right) \omega^1 + \left(f_{122} + \frac{1}{2} f_1K\right) \omega^2, \\ \mathrm{d}f_{22} + 2f_{12}\omega_1{}^2 &= \left(f_{122} - \frac{1}{2} f_1K\right) \omega^1 + f_{222}\omega^2. \end{aligned} \tag{3.19}$$

Definition 3.1. *The invariants*

$$I_{\mathrm{d}f} = f_{11} + f_{22}, \quad J_{\mathrm{d}f} + fI_{\mathrm{d}f} + f^2 = f_{11}f_{22} - f_{12}^2 + f(f_{11} + f_{22} + f) \tag{3.20}$$

are called the Laplacian *and* Weingarten operator *of f and denoted by Δf and $\mathscr{W} f$ resp.*

Now, let us study, in a similar way, a quadratic differential form

$$\begin{aligned} Q &= S_{11}(\omega^1)^2 + 2S_{12}\omega^1\omega^2 + S_{22}(\omega^2)^2 \\ &= {}^*S_{11}({}^*\omega^1)^2 + 2{}^*S_{12}{}^*\omega^1{}^*\omega^2 + {}^*S_{22}({}^*\omega^2)^2 \end{aligned} \tag{3.21}$$

on M. The covariant derivatives $S_{ijk} = S_{jik}$ of S_{ij} be introduced by

$$\begin{aligned} \mathrm{d}S_{11} - 2S_{12}\omega_1{}^2 &= S_{111}\omega^1 + S_{112}\omega^2, \\ \mathrm{d}S_{12} + (S_{11} - S_{22})\, \omega_1{}^2 &= S_{121}\omega^1 + S_{122}\omega^2, \\ \mathrm{d}S_{22} + 2S_{12}\omega_1{}^2 &= S_{221}\omega^1 + S_{222}\omega^2. \end{aligned} \tag{3.22}$$

We get

$$(3.23)\quad \begin{aligned} {}^*S_{11} &= \cos^2\varphi \cdot S_{11} - \sin 2\varphi \cdot S_{12} + \sin^2\varphi \cdot S_{22}, \\ {}^*S_{12} &= \frac{1}{2}\,\varepsilon_1(\sin 2\varphi \cdot S_{11} + 2\cos 2\varphi \cdot S_{12} - \sin 2\varphi \cdot S_{22}), \\ {}^*S_{22} &= \sin^2\varphi \cdot S_{11} + \sin 2\varphi \cdot S_{12} + \cos^2\varphi \cdot S_{22}; \end{aligned}$$

$$(3.24)\quad \begin{aligned} {}^*S_{111} &= \varepsilon_1(\cos^3\varphi \cdot S_{111} - \sin\varphi\cos^2\varphi \cdot S_{112} - 2\sin\varphi\cos^2\varphi \cdot S_{121} \\ &\quad + 2\sin^2\varphi\cos\varphi \cdot S_{122} + \sin^2\varphi\cos\varphi \cdot S_{221} - \sin^3\varphi \cdot S_{222}), \\ {}^*S_{112} &= \sin\varphi\cos^2\varphi \cdot S_{111} + \cos^3\varphi \cdot S_{112} - 2\sin^2\varphi\cos\varphi \cdot S_{121} \\ &\quad - 2\sin\varphi\cos^2\varphi \cdot S_{122} + \sin^3\varphi \cdot S_{221} + \sin^2\varphi\cos\varphi \cdot S_{222}, \\ {}^*S_{121} &= \sin\varphi\cos^2\varphi \cdot S_{111} - \sin^2\varphi\cos\varphi \cdot S_{112} + \cos\varphi\cos 2\varphi \cdot S_{121} \\ &\quad - \sin\varphi\cos 2\varphi \cdot S_{122} - \sin\varphi\cos^2\varphi \cdot S_{221} + \sin^2\varphi\cos\varphi \cdot S_{222}, \\ {}^*S_{122} &= \varepsilon_1(\sin^2\varphi\cos\varphi \cdot S_{111} + \sin\varphi\cos^2\varphi \cdot S_{112} + \sin\varphi\cos 2\varphi \cdot S_{121} \\ &\quad + \cos\varphi\cos 2\varphi \cdot S_{122} - \sin^2\varphi\cos\varphi \cdot S_{221} - \sin\varphi\cos^2\varphi \cdot S_{222}), \\ {}^*S_{221} &= \varepsilon_1(\sin^2\varphi\cos\varphi \cdot S_{111} - \sin^3\varphi \cdot S_{112} + 2\sin\varphi\cos^2\varphi \cdot S_{121} \\ &\quad - 2\sin^2\varphi\cos\varphi \cdot S_{122} + \cos^3\varphi \cdot S_{221} - \sin\varphi\cos^2\varphi \cdot S_{222}), \\ {}^*S_{222} &= \sin^3\varphi \cdot S_{111} + \sin^2\varphi\cos\varphi \cdot S_{112} + 2\sin^2\varphi\cos\varphi \cdot S_{121} \\ &\quad + 2\sin\varphi\cos^2\varphi \cdot S_{122} + \sin\varphi\cos^2\varphi \cdot S_{221} + \cos^3\varphi \cdot S_{222}. \end{aligned}$$

By means of this formulas, it is possible to prove

Lemma 3.4. *To the quadratic form Q on M, associate the functions*

$$(3.25)\quad \begin{aligned} &G_Q = S_{11} + S_{22}, \qquad I_Q = S_{11}S_{22} - S_{12}^2, \\ &J_Q = S_{121}(S_{112} - S_{222}) + S_{122}(S_{221} - S_{111}) \end{aligned}$$

and the 1-form

$$(3.26)\quad \begin{aligned} \hat{Q} &= \{(S_{11} - S_{22})\,S_{121} + S_{12}(S_{221} - S_{111})\}\,\omega^1 \\ &\quad + \{(S_{11} - S_{22})\,S_{122} + S_{12}(S_{222} - S_{112})\}\,\omega^2. \end{aligned}$$

Then

$$(3.27)\quad {}^*G_Q = G_Q, \qquad {}^*I_Q = I_Q, \qquad {}^*J_Q = J_Q$$

and

$$(3.28)\quad {}^*\hat{Q} = \varepsilon_1 Q.$$

From (3.22),

$$(3.29)\quad \begin{aligned} &\{\mathrm{d}S_{111} - (S_{112} + 2S_{121})\,\omega_1{}^2\} \wedge \omega^1 + \{\mathrm{d}S_{112} + (S_{111} - 2S_{122})\,\omega_1{}^2\}\,\omega^2 \\ &= 2S_{12}K\omega^1 \wedge \omega^2, \\ &\{\mathrm{d}S_{121} + (S_{111} - S_{122} - S_{221})\,\omega_1{}^2\} \wedge \omega^1 \\ &+ \{\mathrm{d}S_{122} + (S_{112} + S_{121} - S_{222})\,\omega_1{}^2\} \wedge \omega^2 = (S_{22} - S_{11})\,K\omega^1 \wedge \omega^2, \\ &\{\mathrm{d}S_{221} + (2S_{121} - S_{222})\,\omega_1{}^2 \wedge \omega^1 + \{\mathrm{d}S_{222} + (2S_{122} + S_{221})\,\omega_1{}^2\} \wedge \omega^2 \\ &= -2S_{12}K\omega^1 \wedge \omega^2, \end{aligned}$$

and we get the existence of functions $S_{ijkl} = S_{ijlk}$ such that

$$
\begin{aligned}
(3.30)\quad & \mathrm{d}S_{111} - (S_{112} + 2S_{121})\,\omega_1{}^2 = S_{1111}\omega^1 + (S_{1112} - S_{12}K)\,\omega^2, \\
& \mathrm{d}S_{112} + (S_{111} - 2S_{122})\,\omega_1{}^2 = (S_{1112} + S_{12}K)\,\omega^1 + S_{1122}\omega^2, \\
& \mathrm{d}S_{121} + (S_{111} - S_{122} - S_{221})\,\omega_1{}^2 = S_{1211}\omega^1 + (S_{1212} + S_{11}K)\,\omega^2, \\
& \mathrm{d}S_{122} + (S_{112} - S_{121} - S_{222})\,\omega_1{}^2 = (S_{1212} + S_{22}K)\,\omega^1 + S_{1222}\omega^2, \\
& \mathrm{d}S_{221} + (2S_{121} - S_{222})\,\omega_1{}^2 = S_{2211}\omega^1 + (S_{2212} + S_{12}K)\,\omega^2, \\
& \mathrm{d}S_{222} + (2S_{122} + S_{221})\,\omega_1{}^2 = (S_{2212} - S_{12}K)\,\omega^1 + S_{2222}\omega^2.
\end{aligned}
$$

The proof of the following lemma is easy.

Lemma 3.5. *We have*

$$(3.31)\quad \mathrm{d}\hat{Q} = -\{2J_Q + (S_{11} - S_{22})^2\,K + 4S_{12}^2K\}\,\omega^1 \wedge \omega^2,$$

i.e., $\mathrm{d}\hat{Q}$ *does not depend on the functions* S_{ijkl}.

Finally, let us return to our surface $M\colon \mathscr{D} \to E^3$. Suppose that we have chosen and fixed a unit normal vector field. Then $\varepsilon_2 = 1$ and we may apply the preceding considerations to $Q = II$. Thus — see (2.22) and (2.24) —

$$
\begin{aligned}
(3.32)\quad & S_{11} = a, \quad S_{12} = b, \quad S_{22} = c; \\
& S_{111} = \alpha, \quad S_{112} = S_{121} = \beta, \quad S_{122} = S_{221} = \gamma, \quad S_{222} = \delta; \\
& S_{1111} = A, \quad S_{1112} = B, \quad S_{1122} = C + aK, \quad S_{1211} = B + bK, \quad S_{1212} = C, \\
& S_{1222} = D + bK, \quad S_{2211} = C + cK, \quad S_{2212} = D, \quad S_{2222} = E.
\end{aligned}
$$

From (3.24), we get the promised transformation equations for the quantities $\alpha, \ldots, \delta$:

$$
\begin{aligned}
(3.33)\quad {}^*\alpha &= \varepsilon_1(\cos^3\varphi\cdot\alpha - 3\sin\varphi\cos^2\varphi\cdot\beta + 3\sin^2\varphi\cos\varphi\cdot\gamma - \sin^3\varphi\cdot\delta), \\
{}^*\beta &= \sin\varphi\cos^2\varphi\cdot\alpha + (\cos^3\varphi - 2\sin^2\varphi\cos\varphi)\,\beta \\
&\quad + (\sin^3\varphi - 2\sin\varphi\cos^2\varphi)\,\gamma + \sin^2\varphi\cos\varphi\cdot\delta, \\
{}^*\gamma &= \varepsilon_1\{\sin^2\varphi\cos\varphi\cdot\alpha + (2\sin\varphi\cos^2\varphi - \sin^3\varphi)\,\beta \\
&\quad + (\cos^3\varphi - 2\sin^2\varphi\cos\varphi)\,\gamma - \sin\varphi\cos^2\varphi\cdot\delta\}, \\
{}^*\delta &= \sin^3\varphi\cdot\alpha + 3\sin^2\varphi\cos\varphi\cdot\beta + 3\sin\varphi\cos^2\varphi\cdot\gamma + \cos^3\varphi\cdot\delta.
\end{aligned}
$$

Of course, see (3.25),

$$(3.34)\quad G_{II} = 2H, \quad I_{II} = K,$$

$$(3.35)\quad J_{II} = \beta^2 + \gamma^2 - \beta\delta - \alpha\gamma.$$

Further, see (3.26) and (3.31),

$$(3.36)\quad \widehat{II} = \{(a - c)\,\beta + b(\gamma - \alpha)\}\,\omega^1 + \{(a - c)\,\gamma + b(\delta - \beta)\}\,\omega^2,$$

$$(3.37)\quad \mathrm{d}\widehat{II} = -2\{\beta^2 + \gamma^2 - \beta\delta - \alpha\gamma + 2(H^2 - K)\,K\}\,\omega^1 \wedge \omega^2.$$

V. Global differential geometry of Weingarten surfaces

1. Techniques

Let $M: \mathscr{D} \to E^3$ be a surface. Our main aim will be to prove theorems of the following type: Let I be a certain invariant of M, and let M satisfy the conditions $(\mathscr{P})$ in $\mathscr{D}$ and the conditions $(\mathscr{P}_0)$ on $\partial\mathscr{D}$; then $I = 0$ in $\mathscr{D}$. I am going to show the general procedures for proving such results.

A. Integral formulas

Definition 1.1. *Let $\mathscr{V}$ be a two-dimensional manifold and $M: \mathscr{V} \to E^3$ a surface. Let $U_1, U_2 \subset \mathscr{V}$ be two domains with non-empty intersection and $\{m, v_1, v_2, v_3\}$, $\{m, {}^*v_1, {}^*v_2, {}^*v_3\}$ fields of tangent frames in $M(U_1)$ and $M(U_2)$ resp. These fields are said to be* equally oriented *if we have* $(\text{IV}.2.6)_{\varepsilon_1=1}$ *in $M(U_1 \cap U_2)$. The surface M is called* orientable *if $\mathscr{V}$ may be covered with domains U_ι and we may choose, for each $M(U_\iota)$, a field of frames in such a way that two fields associated to any couple of domains U_ι, $U_{\iota'}$ with $U_\iota \cap U_{\iota'} \neq \emptyset$ are equally oriented.*

Of course, the surface $M: \mathscr{D} \to E^3$ with $\mathscr{D} \subset \mathbf{R}^2$ a domain is always orientable. It is obvious what is meant by the *orientation* of an orientable surface $M: \mathscr{V} \to E^3$; we just choose one of the possible classes of equally orientable fields of frames on M. In what follows, I am going to restrict myself to $\mathscr{V} = \mathscr{D}$ a bounded domain in $\mathbf{R}^2$; nevertheless, *each theorem will be valid if we replace $\mathscr{D}$ by an orientable compact two-dimensional manifold.*

Let us take a (orientable) surface $M: \mathscr{D} \to E^3$, and let us choose one of its orientations. Let ω be an invariant 1-form on (the oriented surface) M in the following sense. Let $\{m, v_1, v_2, v_3\}$, $\{m, {}^*v_1, {}^*v_2, {}^*v_3\}$ be any two fields (possibly not equally oriented) of frames on M connected by (IV.2.6). Let ω be expressed as $R_1\omega^1 + R_2\omega^2$ or ${}^*R_1{}^*\omega^1 + {}^*R_2{}^*\omega^2$ resp.; then ${}^*R_1{}^*\omega^1 + {}^*R_2{}^*\omega^2 = R_1\omega^1 + R_2\omega^2$ or ${}^*R_1{}^*\omega^1 + {}^*R_2{}^*\omega^2 = \varepsilon_1(R_1\omega^1 + R_2\omega^2)$. Writing $\mathrm{d}\omega = J\,\mathrm{d}o$, J is an invariant on the oriented surface M; because of (2.16), we have ${}^*J = \varepsilon_1 J$ or ${}^*J = J$ on the surface itself. The Stokes' Theorem III.1.2 then implies the following

Theorem 1.1. *Let $M: \mathscr{D} \to E^3$ be an oriented surface and ω an invariant 1-form on M satisfying:* (i) $\mathrm{d}\omega = (J_0 + J_1 I)\,\mathrm{d}o$, *$J_0$ and J_1 being invariants on M;* (ii) *the conditions $(\mathscr{P})$ imply $J_0 \geqq 0$, $I \geqq 0$ and $J_1 > 0$ on M;* (iii) *the conditions $(\mathscr{P}_0)$ imply $\omega = 0$ on ∂M. Then $I = 0$ on M.*

Proof follows immediately from

$$0 = \int_{\partial M} \omega = \int_M (J_0 + J_1 I)\,\mathrm{d}o.$$

QED.

As mentioned above, the same theorem holds true for any orientable surface $M\colon \mathscr{V} \to E^3$. If $\partial\mathscr{V} = \emptyset$, we need not worry ourselves with (iii) because of $\int_{\partial\mathscr{V}} \omega = 0$ for any form ω.

How to get appropriate forms ω? Well, the best "procedure" is to guess them. Nevertheless, they are more systematic procedures. One of them is to start with an invariant 1-form φ and consider the 1-form $\omega = *\,\varphi$ or $\omega = \tilde{\varphi}$ resp.; see Lemma IV.3.1 and (IV.3.13) for the definitions. The other one is to construct a 1-form $\tilde{Q}$ from an invariant quadratic form Q as decribed in Lemma IV.3.4. Finally, the most used procedure is to be described as follows. Take an invariant j on M and define $\omega = *\,\mathrm{d}j$. With respect to a chosen field of frames, $\mathrm{d}j = j_1\omega^1 + j_2\omega^2$, $*\,\mathrm{d}j = -j_2\omega^1 + j_1\omega^2$, and, because of equations analogous to (IV.3.17),

$$\mathrm{d} * \mathrm{d}j = (j_{11} + j_{22})\,\omega^1 \wedge \omega^2, \tag{1.1}$$

and the Stokes' theorem reads, see Definition IV.3.1,

$$\int_{\partial M} (-j_2\omega^1 + j_1\omega^2) = \int_M \Delta j \cdot \mathrm{d}o. \tag{1.2}$$

Thus $j_1 = j_2 = 0$ on ∂M and $\Delta_j \geqq 0$ imply $\Delta_j = 0$. There remains "just" one question: how to choose good invariants, 1-forms, quadratic forms, etc., in order to get suitable 1-forms ω? The answer is very simple: to guess them.

B. Maximum principle

In $\mathscr{D}$, let us choose local coordinates (u, v) such that the vectors $\mathrm{d}M(\partial/\partial u)$, $\mathrm{d}M(\partial/\partial v)$ are orthogonal for each point of $\mathscr{D}$; such coordinates do exist. On M, the frames be chosen in such a way that

$$\mathrm{d}M\left(\frac{\partial}{\partial u}\right) = rv_1, \quad \mathrm{d}M\left(\frac{\partial}{\partial v}\right) = sv_2; \quad r = r(u, v) > 0, \quad s = s(u, v) > 0. \tag{1.3}$$

Then

$$\omega^1 = r\,\mathrm{d}u, \qquad \omega^2 = s\,\mathrm{d}v. \tag{1.4}$$

Let $\omega_1{}^2 = \varkappa_1\,\mathrm{d}u + \varkappa_2\,\mathrm{d}v$. Then (IV.2.3) imply

$$\frac{\partial r}{\partial v}\,\mathrm{d}v \wedge \mathrm{d}u = -s\,\mathrm{d}v \wedge \varkappa_1\,\mathrm{d}u, \quad \frac{\partial s}{\partial u}\,\mathrm{d}u \wedge \mathrm{d}v = r\,\mathrm{d}u \wedge \varkappa_2\,\mathrm{d}v,$$

and we get

$$\omega_1{}^2 = -\frac{1}{s}\cdot\frac{\partial r}{\partial v}\,\mathrm{d}u + \frac{1}{r}\cdot\frac{\partial s}{\partial u}\,\mathrm{d}v. \tag{1.5}$$

Now, let f be a function on M; let us pay attention to equations (IV.3.15—16). It is easy to see that

$$f_1 = \frac{1}{r}\,\frac{\partial f}{\partial u}, \qquad f_2 = \frac{1}{s}\,\frac{\partial f}{\partial v}; \tag{1.6}$$

$$f_{11} = \frac{1}{r^2}\,\frac{\partial^2 f}{\partial u^2} - \frac{1}{r^3}\,\frac{\partial r}{\partial u}\,\frac{\partial f}{\partial u} + \frac{1}{rs^2}\,\frac{\partial r}{\partial v}\,\frac{\partial f}{\partial v}, \tag{1.7}$$

$$f_{12} = \frac{1}{rs}\frac{\partial^2 f}{\partial u\,\partial v} - \frac{1}{r^2 s}\frac{\partial r}{\partial v}\frac{\partial f}{\partial u} - \frac{1}{rs^2}\frac{\partial s}{\partial u}\frac{\partial f}{\partial v},$$

$$f_{22} = \frac{1}{s^2}\frac{\partial^2 f}{\partial v^2} + \frac{1}{r^2 s}\frac{\partial s}{\partial u}\frac{\partial f}{\partial u} - \frac{1}{s^3}\frac{\partial s}{\partial v}\frac{\partial f}{\partial v}.$$

Consider an equation of the type

$$b_{11}f_{11} + 2b_{12}f_{12} + b_{22}f_{22} + b_1 f_1 + b_2 f_2 + b_0 f = b. \tag{1.8}$$

Then

$$\begin{aligned}
&\frac{1}{r^2}\,b_{11}\,\frac{\partial^2 f}{\partial u^2} + \frac{2}{rs}\,b_{12}\,\frac{\partial^2 f}{\partial u\,\partial v} + \frac{1}{s^2}\,b_{22}\,\frac{\partial^2 f}{\partial v^2} \\
&+ \left(\frac{1}{r}\,b_1 - \frac{1}{r^3}\frac{\partial r}{\partial u}\,b_{11} - \frac{2}{r^2 s}\frac{\partial r}{\partial v}\,b_{12} + \frac{1}{r^2 s}\frac{\partial s}{\partial u}\,b_{22}\right)\frac{\partial f}{\partial u} \\
&+ \left(\frac{1}{s}\,b_2 + \frac{1}{rs^2}\frac{\partial r}{\partial v}\,b_{11} - \frac{2}{rs^2}\frac{\partial s}{\partial u}\,b_{12} - \frac{1}{s^3}\frac{\partial s}{\partial v}\,b_{22}\right)\frac{\partial f}{\partial v} + b_0 f = b.
\end{aligned} \tag{1.9}$$

Thus we get the following

Theorem 1.2. *Let $M: \mathscr{D} \to E^3$ be a surface with a chosen field of frames. On M, be given functions f, b, b_0, b_i, $b_{ij} = b_{ji}$; f_i and f_{ij} be the covariant derivatives of f with respect to the considered field of frames. Suppose:* (i) *$b_0 \leqq 0$ and $b \geqq 0$ on M;* (ii) *the form $\sum_{i,j=1,2} b_{ij}\xi^i\xi^j$ is definite positive at each point $m \in M$;* (iii) *f satisfies the differential equation* (1.8); (iv) *$f \geqq 0$ on M;* (v) *$f = 0$ on ∂M. Then $f = 0$ on M.*

This theorem holds true even if replacing $\mathscr{D}$ by any two-dimensional manifold $\mathscr{V}$; see Theorem III.2.2.

Finally, let us present (without proof) the following

Theorem 1.3. *Let $M: \mathscr{D} \to E^3$, $\mathscr{D} \subset \mathbf{R}^2$ a bounded domain, be a surface. Then $\mathscr{D}$ may be covered by a single system of local coordinates (u, v) such that*

$$I = r^2\,(\mathrm{d}u^2 + \mathrm{d}v^2), \qquad r = r(u, v) > 0. \tag{1.10}$$

The proof of this (global) theorem is quite complicated and is in connection with the theory of pseudoanalytic functions. The coordinates (u, v) with the property (1.10) are called the *isothermic coordinates*.

Let $\{m, v_1, v_2, v_3\}$ be frames attached to M such that v_1 are tangent to the curves $v = \text{const}$ and v_2 tangent to the curves $u = \text{const}$ resp. Then it is easy to see that we have

$$\omega^1 = r\,\mathrm{d}u, \qquad \omega^2 = r\,\mathrm{d}v. \tag{1.11}$$

Speaking more precisely: taking

$$v_1 = \frac{1}{r}\,(\mathrm{d}M)\left(\frac{\partial}{\partial u}\right), \qquad v_2 = \frac{1}{r}\,(\mathrm{d}M)\left(\frac{\partial}{\partial v}\right), \tag{1.12}$$

it is easy to see that $\langle v_1, v_2\rangle = 0$, $\langle v_1, v_1\rangle = \langle v_2, v_2\rangle = 1$ and we have (1.11).

C. Pseudoanalytic functions

The applications of Theorems III.3.1 and III.3.2 are straightforward and will be made clear in concrete situations.

2. Weingarten surfaces — integral formulas

Definition 2.1. *A surface* $M : \mathscr{D} \to E^3$ *is called a* Weingarten surface *if there is a function* $f : \mathscr{D}' \to \mathbf{R}$, $\mathscr{D}' \subset \mathbf{R}^2$ *being a domain containing* $H(\mathscr{D}) \times K(\mathscr{D})$, *such that* $f(H, K) = 0$ *for each point* $m \in M$. *As above,* H *and* K *are the mean and Gauss curvatures of* M *resp.*

The purpose of this paragraph is to prove, by means of suitable integral formulas, that wide classes of Weingarten surfaces are necessarily parts of spheres.

Theorem 2.1. *Let* $M : \mathscr{D} \to E^3$ *be a surface satisfying:* (i) $K > 0$ *on* M; (ii) *there is a function* $f : \mathscr{D}' \to \mathbf{R}$, $\mathscr{D}' \subset \mathbf{R}^2$ *being a domain containing* $H(\mathscr{D}) \times K(\mathscr{D})$, *such that*

$$(2.1) \qquad f(H, K) = 0, \quad \left(\frac{\partial f}{\partial H}\right)^2 + 4H \frac{\partial f}{\partial H}\frac{\partial f}{\partial K} + 4K\left(\frac{\partial f}{\partial K}\right)^2 > 0$$

on M; (iii) $\mathrm{d}M$ *consists of umbilical points. Then* M *is a part of a sphere.*

Proof. On M, let us consider the 1-form $\widehat{II}$; our integral formula is then — see (IV.3.36) and (IV.3.37) —

$$(2.2) \qquad \int_{\partial M} \{[(a - c)\beta + b(\gamma - \alpha)]\,\omega^1 + [(a - c)\gamma + b(\delta - \beta)]\,\omega^2\}$$
$$= -2\int_M \{\beta^2 + \gamma^2 - \beta\delta - \alpha\gamma + 2(H^2 - K)\,K\}\,\omega^1 \wedge \omega^2.$$

The points on ∂M being umbilical, we have $a = b$, $c = 0$ on ∂M; see (IV.2.27). Thus the left-hand side of (2.2) is equal to zero. For our purpose, it is sufficient to prove that — see (IV.3.35) —

$$(2.3) \qquad J_{II} = \beta^2 + \gamma^2 - \beta\delta - \alpha\gamma \geqq 0;$$

indeed, then — in accord with the general procedure described in Theorem 1.1 — $K > 0$ and (2.3) imply $H^2 - K = 0$ and M is a part of a sphere according to Proposition IV.2.1. Because of (IV.2.26) and (i), $H^2 > 0$. Let $m \in M$ be a fixed point. Then it is possible to choose the field of frames of M in such a way that $H > 0$ on M and $v_1(m)$, $v_2(m)$ are the principal vectors at m, i.e., $b(m) = 0$. From (2.1_1),

$$(2.4) \qquad \frac{\partial f}{\partial H}\,\mathrm{d}H + \frac{\partial f}{\partial K}\,\mathrm{d}K = 0.$$

From (IV.2.19) and (IV.2.22), we get

$$(2.5) \qquad \mathrm{d}H = \frac{1}{2}(\alpha + \gamma)\,\omega^1 + \frac{1}{2}(\beta + \delta)\,\omega^2, \quad \mathrm{d}K = (a\gamma + c\alpha)\,\omega^1 + (a\delta + c\beta)\,\omega^2$$

at m.

Thus (2.4) yields

$$\frac{\partial f}{\partial H}(\alpha+\gamma)+2\frac{\partial f}{\partial K}(a\gamma+c\alpha)=0,$$

$$\frac{\partial f}{\partial H}(\beta+\delta)+2\frac{\partial f}{\partial K}(a\delta+c\beta)=0 \quad \text{at } m,$$

i.e.,

$$\left(\frac{\partial f}{\partial H}+2c\frac{\partial f}{\partial K}\right)\alpha+\left(\frac{\partial f}{\partial H}+2a\frac{\partial f}{\partial K}\right)\gamma=0, \tag{2.6}$$

$$\left(\frac{\partial f}{\partial H}+2c\frac{\partial f}{\partial K}\right)\beta+\left(\frac{\partial f}{\partial H}+2a\frac{\partial f}{\partial K}\right)\delta=0 \quad \text{at } m.$$

This and (2.1_2) implies the existence of numbers ϱ, σ such that

$$\alpha=\varrho\left(\frac{\partial f}{\partial H}+2a\frac{\partial f}{\partial K}\right),\quad \gamma=-\varrho\left(\frac{\partial f}{\partial H}+2c\frac{\partial f}{\partial K}\right), \tag{2.7}$$

$$\beta=\sigma\left(\frac{\partial f}{\partial H}+2a\frac{\partial f}{\partial K}\right),\quad \delta=-\sigma\left(\frac{\partial f}{\partial H}+2c\frac{\partial f}{\partial K}\right) \quad \text{at } m,$$

and we get — see (2.3) —

$$J_{II}(m)=\beta^2+\gamma^2+(\sigma^2+\varrho^2)\left\{\left(\frac{\partial f}{\partial H}\right)^2+4H\frac{\partial f}{\partial H}\frac{\partial f}{\partial K}+4K\left(\frac{\partial f}{\partial K}\right)^2\right\}\geqq 0. \tag{2.8}$$

QED.

This theorem implies two famous corollaries:

Corollary 2.1 (H-theorem). *Let $M:\mathscr{D}\to E^3$ be a surface satisfying:* (i) $K>0$ *on* M; (ii) $H=\text{const}$ *on* M; (iii) ∂M *consists of umbilical points. Then M is a part of a sphere.*

Corollary 2.2 (K-theorem). *Let $M:\mathscr{D}\to E^3$ be a surface satisfying:* (i) $K=\text{const}>0$ *on* M; (ii) ∂M *consists of umbilical points. Then M is a part of a sphere.*

For the proof, take $F(H,K)=H-\text{const}$ or $F(H,K)=K-\text{const}$ resp.

The next theorem is a generalization of the H-theorem: indeed, $H=\text{const}$ implies $H_1=H_2=0$ and this together with $K>0$ imply (i) and (ii). Notice that $H_1{}^2+H_2{}^2$ is an invariant according to Lemma 3.1: we have $H_1{}^2+H_2{}^2=G_{\mathrm{d}H}$.

Theorem 2.2. *Let M be a surface satisfying:* (i) *there is no open domain* $M_1\subset M$ *such that* $K=0$ *on* M_1; (ii) *we have*

$$4(H^2-K)K\geqq H_1{}^2+H_2{}^2 \tag{2.9}$$

on M; (iii) ∂M *consists of umbilical points. Then M is a part of a sphere.*

Proof. Our starting point is again the integral formula (2.2); because of (iii), its left-hand side is equal to zero. From (IV.2.22),

$$\mathrm{d}H=H_1\omega^1+H_2\omega^2=\frac{1}{2}(\alpha+\gamma)\,\omega^1+\frac{1}{2}(\beta+\delta)\,\omega^2, \tag{2.10}$$

i.e.,

$$(2.11)\qquad H_1 = \frac{1}{2}(\alpha+\gamma),\quad H_2 = \frac{1}{2}(\beta+\delta).$$

It is easy to see that (2.2) may be given the form

$$(2.12)\qquad \int_M \{(2\beta - H_2)^2 + (2\gamma - H_1)^2 + 4(H^2 - K)\,K - H_1^2 - H_2^2\}\,\omega^1 \wedge \omega^2 = 0;$$

(2.9) implies $2\beta = H_2$, $2\gamma = H_1$, i.e.,

$$(2.13)\qquad \alpha = 3\gamma,\qquad \delta = 3\beta$$

and

$$(2.14)\qquad 4(H^2 - K)\,K = H_1^2 + H_2^2.$$

Suppose that $m \in M$ is a non-umbilical point, i.e., $H^2 - K > 0$ at m. In a possibly small neighbourhood U of this point m, we may choose the frames in such a way that $b = 0$, $a \neq c$ i.e., v_1 and v_2 are, in U, the principal vectors. Let us restrict ourselves to U. The equations (IV.2.22) reduce to

$$(2.15)\qquad \mathrm{d}a = 3\gamma\omega^1 + \beta\omega^2,\quad (a-c)\,\omega_1^2 = \beta\omega^1 + \gamma\omega^2,\quad \mathrm{d}c = \gamma\omega^1 + 3\beta\omega^2$$

because of (2.13). By exterior differentiation, taking regard of (2.15_2),

$$(2.16)\qquad 3(\mathrm{d}\gamma - \beta\omega_1^2)\wedge\omega^1 + (\mathrm{d}\beta + \gamma\omega_1^2)\wedge\omega^2 = 0,$$

$$(\mathrm{d}\beta + \gamma\omega_1^2)\wedge\omega^1 + (\mathrm{d}\gamma - \beta\omega_1^2)\wedge\omega^2 = (c-a)\,ac\omega^1\wedge\omega^2,$$

$$(\mathrm{d}\gamma - \beta\omega_1^2)\wedge\omega^1 + 3(\mathrm{d}\beta + \gamma\omega_1^2)\wedge\omega^2 = 0.$$

Thus there is a function $\varrho\colon U \to \mathbf{R}$ such that

$$(2.17)\qquad \mathrm{d}\beta + \gamma\omega_1^2 = (\varrho - ac^2)\,\omega^2,\qquad \mathrm{d}\gamma - \beta\omega_1^2 = (\varrho - a^2c)\,\omega^1$$

as a consequence of the Cartan's lemma. By exterior differentiation,

$$(2.18)\qquad (\mathrm{d}\varrho - 3a^2\beta\omega^2)\wedge\omega^1 = 0,\qquad (\mathrm{d}\varrho - 3c^2\gamma\omega^1)\wedge\omega^2 = 0,$$

i.e.,

$$(2.19)\qquad \mathrm{d}\varrho = 3c^2\gamma\omega^1 + 3a^2\beta\omega^2.$$

A further exterior differentiation yields

$$(2.20)\qquad (a-c)\,\beta\gamma = 0.$$

Suppose $\beta = 0 \neq \gamma$, the case $\beta \neq 0 = \gamma$ being symmetric. We have

$$(2.21)\qquad \mathrm{d}a = 3\gamma\omega^1,\qquad \mathrm{d}c = \gamma\omega^1,$$

and the equation (2.14) turns out to be

$$(2.22)\qquad a^3c - 2a^2c^2 + ac^3 = 4\gamma^2.$$

We get (taking regard of $\gamma \neq 0$!)

$$(2.23)\qquad a^3 + 13a^2c - 9ac^2 + 3c^3 = 8\varrho,\quad 11a^2 + 30ac - 21c^2 = 0,\quad 2a + c = 0$$

by a series of successive differentiations. The differentiation of (2.23_3) implies $\gamma = 0$, hence a contradiction. Thus we have, in U, $\beta = \gamma = 0$, $\omega_1{}^2 = 0$ and $a \neq c$. The exterior differentiation of $\omega_1{}^2 = 0$ yields $ac\omega^1 \wedge \omega^2 = 0$, i.e., $K = 0$ in U. This being a contradiction to (i), m cannot be a non-umbilical point. Thus all points of M are umbilical and M is not a part of a plane because of (i). QED.

Now, let us try to characterize a class of surfaces with constant mean curvature.

Theorem 2.3. *Be given a surface* $M: \mathscr{D} \to E^3$ *satisfying:* (i) $K > 0$ *on* M; (ii) *on* M, *there is a couple of orthogonal unit tangent vector fields* V_1, V_2 *such that*

$$(2.24) \qquad V_1 V_1 H = 0, \qquad V_2 H = 0;$$

(iii) *on* ∂M,

$$(2.25) \qquad V_1 H = 0, \qquad V_2 H = 0.$$

Then $H =$ const *on* M.

Proof. The tangent frames associated to M may be chosen in such a way that

$$(2.26) \qquad V_1 = v_1, \qquad V_2 = v_2.$$

The condition (2.24_2) reads, see (2.11),

$$(2.27) \qquad \beta + \delta = 0$$

on M. Further, see (IV.2.24) and take regard of (2.27),

$$(2.28) \qquad \mathrm{d}(v_1 H) = \frac{1}{2}\,\mathrm{d}(\alpha + \gamma) = \frac{1}{2}\,(A + C + cK)\,\omega^1 + \frac{1}{2}\,(B + D)\,\omega^2,$$

and the condition (2.24) may be written as

$$(2.29) \qquad A + C + cK = 0$$

on M. Of course, (iii) reads

$$(2.30) \qquad \alpha + \gamma = \beta + \delta = 0 \quad \text{on } \partial M.$$

From (IV.2.24), (2.27) and (2.29),

$$(2.31) \qquad \mathrm{d}(\alpha + \gamma) = (B + D)\,\omega^2, \quad (\alpha + \gamma)\,\omega_1{}^2 = (B + D)\,\omega^1 + (C + E + aK)\,\omega^2.$$

On M, consider the 1-form

$$(2.32) \qquad \varphi = (\alpha + \gamma)^2\,\omega_1{}^2;$$

the frames being fixed, it is invariant. Then

$$(2.32) \qquad \mathrm{d}\varphi = 2(\alpha + \gamma)\,(B + D)\,\omega^2 \wedge \omega_1{}^2 - (\alpha + \gamma)^2\,K\omega^1 \wedge \omega^2$$
$$= 2(B + D)^2\,\omega^2 \wedge \omega^1 - (\alpha + \gamma)^2\,K\omega^1 \wedge \omega^2.$$

Because of (2.30), we have $\varphi = 0$ on ∂M, and the Stokes' formula $\int_{\partial M} \varphi = \int_M \mathrm{d}\varphi$ takes the form

$$\int_M \{2(B+D)^2 + (\alpha+\gamma)^2 K\}\, \omega^1 \wedge \omega^2 = 0. \tag{2.33}$$

Because of the supposition (i), $\alpha + \gamma = 0$ on M and this together with (2.27) implies $dH = 0$ on M. QED.

Let us remark that it may be proved that there are surfaces (at least of class C^ω) possessing two orthogonal unit tangent vector fields V_1, V_2 such that (2.24) is valid without having a constant mean curvature (of course, they do not fulfill the boundary condition (iii) of the preceding theorem).

Combining Corollary 2.1 and Theorem 2.3, we get

Corollary 2.3. *Be given a surface* $M: \mathscr{D} \to E^3$ *satisfying:* (i) $K > 0$ *on* M; (ii) *there is, on* M, *a couple of orthogonal unit tangent vector fields* V_1, V_2 *such that* (2.24) *is valid*; (iii) *the points of* ∂M *are umbilical and satisfy* (2.25). *Then* M *is a part of a sphere.*

In the next paragraph, we shall see that, using the method of pseudoanalytic functions, Theorem 2.1 remains valid even if we remove the condition (i). Thus we are led to the natural question whether by using other (i.e., better) formula than (2.2) we possibly cannot attain this more general result. In the following, I am going to show that this is not the case at least in a certain class of integral formulas. In this context the restriction to surfaces with a net of lines of curvature is not essential.

Let $M: \mathscr{D} \to E^3$ be a surface, ∂M its boundary, and let us suppose that on M there is an orthogonal net of lines of curvature. The moving orthonormal frames $\{m, v_1, v_2, v_3\}$ be chosen in such a way that v_1, v_2 are tangent to the lines of curvature. Then we have the formulas (IV.2.17), (IV.2.22), (IV.2.24) with

$$b = 0. \tag{2.34}$$

The functions a and c are the principal curvatures and, because of (IV.2.22$_{1,3}$),

$$v_1 a = \alpha, \quad v_2 a = \beta, \quad v_1 c = \gamma, \quad v_2 c = \delta. \tag{2.35}$$

Our task is to produce all invariant 1-forms φ on M with the following properties: (i) $\varphi = 0$ in the umbilical points; (ii) both the forms φ and $d\varphi$ depend just on the principal curvatures and theirs derivatives of order one. Notice that $\widehat{II}$ really satisfies both requirements.

Because of (i) and the first part of (ii), φ has to have the form

$$\varphi = f(a, c, \alpha, \beta, \gamma, \delta) \cdot \varphi^1 + g(a, c, \alpha, \beta, \gamma, \delta) \cdot \varphi^2 \tag{2.36}$$

with

$$\varphi^1 = (a-c)\,\omega^1, \quad \varphi^2 = (a-c)\,\omega^2. \tag{2.37}$$

Of course,

$$d\varphi^1 = \delta\omega^1 \wedge \omega^2, \quad d\varphi^2 = \alpha\omega^1 \wedge \omega^2. \tag{2.38}$$

Further, it is just a computational matter to show that

$$(2.39)\quad (a-c)\,\mathrm{d}f = (.)\,\omega^1 + \Big\{(a-c)\Big(aK\frac{\partial f}{\partial\beta} + \frac{\partial f}{\partial a}\beta + \frac{\partial f}{\partial c}\delta + \frac{\partial f}{\partial\alpha}B + \frac{\partial f}{\partial\beta}C + \frac{\partial f}{\partial\gamma}D + \frac{\partial f}{\partial\delta}E\Big) - \frac{\partial f}{\partial\beta}\alpha\gamma + \Big(3\frac{\partial f}{\partial\alpha} - 2\frac{\partial f}{\partial\gamma}\Big)\beta\gamma + \Big(2\frac{\partial f}{\partial\beta} - 3\frac{\partial f}{\partial\delta}\Big)\gamma^2 + \frac{\partial f}{\partial\gamma}\gamma\delta\Big\}\,\omega^2,$$

$$(a-c)\,\mathrm{d}g = \Big\{(a-c)\Big(cK\frac{\partial g}{\partial\gamma} + \frac{\partial g}{\partial a}\alpha + \frac{\partial g}{\partial c}\gamma + \frac{\partial g}{\partial\alpha}A + \frac{\partial g}{\partial\beta}B + \frac{\partial g}{\partial\gamma}C + \frac{\partial g}{\partial\delta}D\Big) - \frac{\partial g}{\partial\beta}\alpha\beta + \Big(3\frac{\partial g}{\partial\alpha} - 2\frac{\partial g}{\partial\gamma}\Big)\beta^2 + \Big(2\frac{\partial g}{\partial\beta} - 3\frac{\partial g}{\partial\delta}\Big)\beta\gamma + \frac{\partial g}{\partial\gamma}\beta\delta\Big\}\,\omega^1 + (.)\,\omega^2$$

and

$$(2.40)\quad \mathrm{d}\varphi = \Big\{(a-c)\Big(K\Big(c\frac{\partial g}{\partial\gamma} - a\frac{\partial f}{\partial\beta}\Big) + \frac{\partial g}{\partial a}\alpha - \frac{\partial f}{\partial a}\beta + \frac{\partial g}{\partial c}\gamma - \frac{\partial f}{\partial c}\delta + \frac{\partial g}{\partial\alpha}A + \Big(\frac{\partial g}{\partial\beta} - \frac{\partial f}{\partial\alpha}\Big)B + \Big(\frac{\partial g}{\partial\gamma} - \frac{\partial f}{\partial\beta}\Big)C + \Big(\frac{\partial g}{\partial\delta} - \frac{\partial f}{\partial\gamma}\Big)D - \frac{\partial f}{\partial\delta}E\Big) - \frac{\partial g}{\partial\beta}\alpha\beta + \frac{\partial f}{\partial\beta}\alpha\gamma + \Big(3\frac{\partial g}{\partial\alpha} - 2\frac{\partial g}{\partial\gamma}\Big)\beta^2 + \Big(2\frac{\partial g}{\partial\beta} - 3\frac{\partial g}{\partial\delta} - 3\frac{\partial f}{\partial\alpha} + 2\frac{\partial f}{\partial\gamma}\Big)\beta\gamma + \Big(3\frac{\partial f}{\partial\delta} - 2\frac{\partial f}{\partial\beta}\Big)\gamma^2 + \frac{\partial g}{\partial\gamma}\beta\delta - \frac{\partial f}{\partial\gamma}\gamma\delta + g\alpha + f\delta\Big\}\,\omega^1\wedge\omega^2.$$

Because of the second part of (ii),

$$(2.41)\quad \frac{\partial g}{\partial\alpha} = 0,\quad \frac{\partial g}{\partial\beta} = \frac{\partial f}{\partial\alpha},\quad \frac{\partial g}{\partial\gamma} = \frac{\partial f}{\partial\beta},\quad \frac{\partial g}{\partial\delta} = \frac{\partial f}{\partial\gamma},\quad 0 = \frac{\partial f}{\partial\delta}.$$

The convenient combinations of the derivatives of the equations (2.41) yield that f, γ are linear in $\alpha, \beta, \gamma, \delta$ and then it is easy to see that the general solution of (2.41) is

$$(2.42)\quad \begin{aligned} f &= R(a,c)\cdot\alpha + S(a,c)\cdot\beta + T(a,c)\cdot\gamma + V(a,c),\\ g &= R(a,c)\cdot\beta + S(a,c)\cdot\gamma + T(a,c)\cdot\delta + W(a,c). \end{aligned}$$

The solution of our problem is thus contained in the following

Lemma 2.1. *Let $M: \mathscr{D} \to E^3$ be a surface with an orthogonal net of lines of curvature; the specialization of frames and the notation be as above. Let φ be an invariant 1-form on M satisfying:* (i) *$\varphi = 0$ in the umbilical points;* (ii) *both the forms φ and $\mathrm{d}\varphi$ depend just on the principal curvatures and theirs first order derivatives. Then φ is* *of the form*

$$(2.43) \qquad \varphi = (a - c)\,(f\omega^1 + g\omega^2),$$

where f and g are given by (2.42). *We have*

$$(2.44) \qquad \mathrm{d}\varphi = \{-(a - c)^2\, SK + \Phi\}\, \omega^1 \wedge \omega^2$$

with

$$(2.45) \qquad \Phi = \left\{2S + (a - c)\frac{\partial S}{\partial a}\right\} \alpha\gamma + \left\{T + R + (a - c)\left(\frac{\partial T}{\partial a} - \frac{\partial R}{\partial c}\right)\right\}(\alpha\delta - \beta\gamma)$$
$$+ \left\{2S - (a - c)\frac{\partial S}{\partial c}\right\}\beta\delta - \left\{2S + (a - c)\frac{\partial S}{\partial a}\right\}\beta^2 - \left\{2S - (a - c)\frac{\partial S}{\partial c}\right\}\gamma^2$$
$$+ \left\{W + (a - c)\frac{\partial W}{\partial a}\right\}\alpha - (a - c)\frac{\partial V}{\partial a}\beta$$
$$+ (a - c)\frac{\partial W}{\partial c}\gamma + \left\{V - (a - c)\frac{\partial V}{\partial c}\right\}\delta.$$

From the Stokes' formula $\int_{\partial M} \varphi = \int_M \mathrm{d}\varphi$, we get the following very general

Theorem 2.4. *Let $M: \mathscr{D} \to E^3$ be a surface with an orthogonal net of lines of curvature, let a, c be its principal curvatures. Suppose:* (i) *there are functions $S(a, c)$, $R(a, c)$, $T(a, c)$, $V(a, c)$, $W(a, c)$ such that $S(a, c)\,K > 0$ and $\Phi \leqq 0$ on M, Φ being given by* (2.45); (ii) *∂M consists of umbilical points. Then M is a part of a sphere.*

Set

$$(2.46) \qquad S = \frac{1}{2}; \qquad T, R, V, W = \text{const}; \qquad P := T + R.$$

Then

$$(2.47) \qquad \Phi = \alpha\gamma + P(\alpha\delta - \beta\gamma) + \beta\delta - \beta^2 - \gamma^2 + W\alpha + V\delta,$$

and we get the following

Corollary 2.4. *Let $M: \mathscr{D} \to E^3$ be a surface with an orthogonal net of lines of curvature satisfying:* (i) *they are numbers $P, W, V \in \mathbf{R}$ such that*

$$(2.48) \qquad P(\alpha\delta - \beta\gamma) + W\alpha + V\delta \leqq \beta^2 + \gamma^2 - \beta\delta - \alpha\gamma$$

on M; (ii) *∂M consists of umbilical points. Then M is a part of a sphere.*

Following the proof of Theorem 2.1, we see that this theorem is contained in Corollary 2.4.

In proving that M is a sphere, we should prove that $a - c = 0$ is a consequence of some conditions imposed on M. But the class of forms φ introduced above is bad in this respect. Indeed, the factor of $(a - c)$ in (2.44) is K, and I can see no way how to get rid of it. Theoretically, one should examine more wide classes of invariant forms φ on M.

3. Weingarten surfaces — pseudoanalytic functions

The Weingarten condition (2.1_1) implies (2.4). We are now going to prove, by means of the theory of pseudoanalytic functions, a theorem concerning surfaces satisfying a generalization of (2.4). Theorem 2.1 will be then a special case of this result, namely, we have to suppose $K > 0$ and

$$R_1 = \frac{\partial f}{\partial H}, \quad R_2 = \frac{\partial f}{\partial K}, \quad R_3 = R_4 = 0. \tag{3.1}$$

The $*$-operator is to be defined by (3.13); see Lemma IV.3.3. On M, we have to take a fixed orientation.

Theorem 3.1. *Let $M: \mathscr{D} \to E^3$ be a surface, $\mathscr{D} \subset \mathbf{R}^2$ a bounded domain. Let M satisfy:* (i) *there are real-valued functions R_1, R_2, R_3, R_4 on M such that*

$$R_1\,\mathrm{d}H + R_2\,\mathrm{d}K + R_3 * \mathrm{d}H + R_4 * \mathrm{d}K = 0, \tag{3.2}$$

$$R_1{}^2 + R_3{}^2 + 4H(R_1R_2 + R_3R_4) + 4K(R_2{}^2 + R_4{}^2) > 0 \tag{3.3}$$

on M; (ii) *∂M consists of umbilical points. Then M is a part of a sphere.*

Proof. We have

$$\mathrm{d}H = \frac{1}{2}(\alpha + \gamma)\,\omega^1 + \frac{1}{2}(\beta + \delta)\,\omega^2, \tag{3.4}$$

$$* \mathrm{d}H = -\frac{1}{2}(\beta + \delta)\,\omega^1 + \frac{1}{2}(\alpha + \gamma)\,\omega^2,$$

$$\mathrm{d}K = (a\gamma - 2b\beta + c\alpha)\,\omega^1 + (a\delta - 2b\gamma + c\beta)\,\omega^2,$$

$$* \mathrm{d}K = (2b\gamma - a\delta - c\beta)\,\omega^1 + (a\gamma - 2b\beta + c\alpha)\,\omega^2,$$

so that the equation (3.2) implies

$$R_1(\alpha + \gamma) + 2R_2(a\gamma - 2b\beta + c\alpha) - R_3(\beta + \delta) - 2R_4(a\delta - 2b\gamma + c\beta) = 0, \tag{3.5}$$

$$R_1(\beta + \delta) + 2R_2(a\delta - 2b\gamma + c\beta) + R_3(\alpha + \gamma) + 2R_4(a\gamma - 2b\beta + c\alpha) = 0.$$

In $\mathscr{D}$, choose an orthogonal coordinate system (u, v) such that we have (1.4) and (1.5). From (IV.2.22),

$$d(a - c) = 4b\omega_1{}^2 + (\alpha - \gamma)\,\omega^1 + (\beta - \delta)\,\omega^2, \tag{3.6}$$

$$\mathrm{d}b = -(a - c)\,\omega_1{}^2 + \beta\omega^1 + \gamma\omega^2,$$

i.e.,

$$\frac{\partial(a - c)}{\partial u} = (\alpha - \gamma)\,r - \frac{4}{s}\frac{\partial r}{\partial v}\,b, \quad \frac{\partial(a - c)}{\partial v} = (\beta - \delta)\,s + \frac{4}{r}\frac{\partial s}{\partial u}\,b, \tag{3.7}$$

$$\frac{\partial b}{\partial u} = \beta r + \frac{1}{s}\frac{\partial r}{\partial v}(a - c), \quad \frac{\partial b}{\partial v} = \gamma s - \frac{1}{r}\frac{\partial s}{\partial u}(a - c).$$

From this,

$$\text{(3.8)} \qquad \alpha rs = s \frac{\partial(a-c)}{\partial u} + r \frac{\partial b}{\partial v} + (.)\,(a-c) + (.)\,b,$$

$$\beta rs = s \frac{\partial b}{\partial u} + (.)\,(a-c) + (.)\,b,$$

$$\gamma rs = r \frac{\partial b}{\partial v} + (.)\,(a-c) + (.)\,b,$$

$$\delta rs = -r \frac{\partial(a-c)}{\partial v} + s \frac{\partial b}{\partial u} + (.)\,(a-c) + (.)\,b.$$

The system (3.5) becomes

$$\text{(3.9)} \qquad a_{11} \frac{\partial(a-c)}{\partial u} + a_{12} \frac{\partial(a-c)}{\partial v} + b_{11} \frac{\partial b}{\partial u} + b_{12} \frac{\partial b}{\partial v} = c_{11}(a-c) + c_{12} b,$$

$$a_{21} \frac{\partial(a-c)}{\partial u} + a_{22} \frac{\partial(a-c)}{\partial v} + b_{21} \frac{\partial b}{\partial u} + b_{22} \frac{\partial b}{\partial v} = c_{21}(a-c) + c_{22} b$$

with

$$\text{(3.10)} \qquad a_{11} = s(R_1 + 2cR_2), \qquad a_{12} = r(R_3 + 2aR_4),$$

$$b_{11} = -2s(2bR_2 + R_3 + 4HR_4), \quad b_{12} = 2r(R_1 + 2HR_2 + 2bR_4),$$

$$a_{21} = s(R_3 + 2cR_4), \qquad a_{22} = -r(R_1 + 2aR_2),$$

$$b_{21} = 2s(R_1 + 2HR_2 - 2bR_4), \qquad b_{22} = 2r(-2bR_2 + R_3 + 2HR_4).$$

We have to calculate the associated form (III.3.2). From (3.10),

$$\text{(3.11)} \qquad a_{12}b_{22} - a_{22}b_{12} = 2r^2[2(H+a)\,(R_1R_2 + R_3R_4) + 2b(R_1R_4 - R_2R_3) + 4aH(R_2^2 + R_4^2) + R_1^2 + R_3^2],$$

$$a_{11}b_{22} - a_{21}b_{12} + a_{12}b_{21} - a_{22}b_{11} = 4rs[-2b(R_1R_2 + R_3R_4) + (a-c)\,(R_1R_4 - R_2R_3) - 4bH(R_2^2 + R_4^2)],$$

$$a_{11}b_{21} - a_{21}b_{11} = 2s^2[R_1^2 + R_2^2 + 2(H+c)\,(R_1R_2 + R_3R_4) + 2b(R_2R_3 - R_1R_4) + 4cH(R_2^2 + R_4^2)].$$

Denoting by $\varDelta$ the discriminant of Φ, we get

$$\text{(3.12)} \qquad \frac{\varDelta}{16r^2s^2} = [(R_1 + 2HR_2)^2 + (R_3 + 2HR_4)^2] \times [R_1^2 + R_3^2 + 4H(R_1R_2 + R_3R_4) + 4K(R_2^2 + R_4^2)].$$

The first term of the right-hand side product cannot be equal to zero; indeed, let us suppose, on contrary, $R_1 + 2HR_2 = R_3 + 2HR_4 = 0$. Then the second term would be equal to $-4(H^2 - K)(R_2^2 + R_4^2) \leqq 0$, which is a contradiction to (3.3). This means that (2.3) induces the system (2.9) to be elliptic. On the boundary ∂M, $a - c = b = 0$ according to (ii). From this, $a - c = b = 0$ on M, i.e., $H^2 - K = 0$ on M. QED.

To show the wide possibilities for the exploitation of the theory of pseudoanalytic functions, let us prove two more results.

Theorem 3.2. *Let* $M: \mathcal{D} \to E^3$, $\mathcal{D} \subset \mathbf{R}^2$ *a bounded domain, be a surface satisfying:* (i) *on* M, *there is a couple of orthogonal unit tangent vector fields* V_1, V_2 *such that*

$$(3.13) \qquad V_1 H = 0, \qquad V_2 K = 0;$$

(ii) *on* M,

$$(3.14) \qquad 2K + 2[II(V_1, V_1)]^2 + [II(V_1, V_2)]^2 > 0,$$

$II(.,.)$ *being the bilinear form associated to the second fundamental form* $II(.)$ *of* M; (iii) ∂M *consists of umbilical points. Then* M *is a part of a sphere.*

Proof. We may choose the frames in such a way that we have (2.26). Because of (2.11) and (3.4_3), (3.13) may be written as

$$(3.15) \qquad \alpha + \gamma = 0, \qquad a\delta - 2b\gamma + c\beta = 0.$$

From (3.8), we get

$$(3.16) \qquad s\,\frac{\partial(a-c)}{\partial u} + 2r\,\frac{\partial b}{\partial v} = (.)\,(a-c) + (.)\,b,$$

$$ar\,\frac{\partial(a-c)}{\partial v} - (a+c)\,s\,\frac{\partial b}{\partial u} + 2br\,\frac{\partial b}{\partial v} = (.)\,(a-c) + (.)\,b.$$

This system has the form (III.3.1), and the associated form (III.3.2) is equal to

$$(3.17) \qquad \Phi = 2ar^2\mu^2 + 2brs\mu\nu + (a+c)\,s^2\nu^2.$$

The discriminant of Φ is equal to

$$(3.18) \qquad \Delta = r^2s^2(2a^2 + 2ac - b^2) = r^2s^2(2a^2 + b^2 + 2K)$$
$$= r^2s^2\{2[II(v_1, v_1)]^2 + [II(v_1, v_2)]^2 + 2K\};$$

the system (3.16) is thus elliptic. Our assertion follows then from Theorem III.3.1. QED.

The next theorem is a generalization of the previous one, but this time the conditions (3.14) are substituted by conditions of other kind.

Theorem 3.3. *Let* $M: \mathcal{D} \to E^3$, $\mathcal{D} \subset \mathbf{R}^2$ *a bounded domain, be a surface satisfying:* (i) *there is a couple of orthonormal tangent vector fields* V_1, V_2 *on* M *and a couple of functions* $f, g: \mathbf{R}^2 \to \mathbf{R}$ *(defined, perhaps, on a smaller domain containing* $H(M) \times K(M)$) *such that*

$$(3.19) \qquad V_1 g(H, K) = 0, \qquad V_2 f(H, K) = 0,$$

$$(3.20) \qquad \frac{\partial g}{\partial H} \cdot \frac{\partial g}{\partial K} > 0, \qquad \frac{\partial f}{\partial H} \cdot \frac{\partial f}{\partial K} > 0;$$

(ii) $K > 0$ *on* M; (iii) ∂M *consists of umbilical points. Then* M *is a part of a sphere.*

Proof. Let us recall the equations (3.4) and (3.7). The equations (3.19) turn out to be

$$\text{(3.21)} \qquad \frac{\partial g}{\partial H}(\alpha + \gamma) + 2\frac{\partial g}{\partial K}(a\gamma + c\alpha - 2b\beta) = 0,$$

$$\frac{\partial f}{\partial H}(\beta + \delta) + 2\frac{\partial f}{\partial K}(a\delta + c\beta - 2b\gamma) = 0,$$

i.e.,

$$\text{(3.22)} \qquad s\left(\frac{\partial g}{\partial H} - 2c\frac{\partial g}{\partial K}\right)\frac{\partial(a-c)}{\partial u} - 4sb\frac{\partial g}{\partial K}\frac{\partial b}{\partial u} + 2r\left(\frac{\partial g}{\partial H} + 2H\frac{\partial g}{\partial K}\right)\frac{\partial b}{\partial v}$$
$$= (.)\,(a-c) + (.)\,b,$$
$$-r\left(\frac{\partial f}{\partial H} + 2a\frac{\partial f}{\partial K}\right)\frac{\partial(a-c)}{\partial v} + 2s\left(\frac{\partial f}{\partial H} + 2H\frac{\partial f}{\partial K}\right)\frac{\partial b}{\partial u} - 4rb\frac{\partial f}{\partial K}\frac{\partial b}{\partial v}$$
$$= (.)\,(a-c) + (.)\,b.$$

The discriminant of the associated form (III.3.2) is equal to

$$\text{(3.23)} \qquad \varDelta = 16r^2s^2\left\{2H\frac{\partial f}{\partial H}\frac{\partial f}{\partial K}\left(\frac{\partial g}{\partial H}\right)^2 + 2H\frac{\partial g}{\partial H}\frac{\partial g}{\partial K}\left(\frac{\partial f}{\partial H}\right)^2\right.$$
$$+ 2\left(\frac{1}{2}\frac{\partial g}{\partial H} + c\frac{\partial g}{\partial K}\right)^2\left(\frac{\partial f}{\partial H}\right)^2 + 2\left(\frac{1}{2}\frac{\partial f}{\partial H} + a\frac{\partial f}{\partial K}\right)^2\left(\frac{\partial g}{\partial H}\right)^2$$
$$+ (2K + b^2)\left(\left(\frac{\partial f}{\partial H}\right)^2\left(\frac{\partial g}{\partial K}\right)^2 + \left(\frac{\partial f}{\partial K}\right)^2\left(\frac{\partial g}{\partial H}\right)^2\right)$$
$$+ 2(6H^2 + 2K + b^2)\frac{\partial f}{\partial H}\frac{\partial f}{\partial K}\frac{\partial g}{\partial H}\frac{\partial g}{\partial K}$$
$$+ 8H(b^2 + c^2 + 2K)\frac{\partial f}{\partial H}\frac{\partial f}{\partial K}\left(\frac{\partial g}{\partial K}\right)^2$$
$$+ 8H(a^2 + b^2 + 2K)\frac{\partial g}{\partial H}\frac{\partial g}{\partial K}\left(\frac{\partial f}{\partial K}\right)^2$$
$$\left. + 16H^2K\left(\frac{\partial f}{\partial K}\right)^2\left(\frac{\partial g}{\partial K}\right)^2\right\}.$$

Because of $K > 0$ and (3.20), $\varDelta > 0$ and the theorem follows. QED.

There are many other ways how to characterize the spheres. Let us show one of them.

Let $M: \mathscr{D} \to E^3$ be a surface and $S \in E^3$ be a fixed point. For each $d \in \mathscr{D}$, define the vector $v(d)$ by

$$\text{(3.24)} \qquad S = M(d) + v(d).$$

Let v_3 be a fixed field of the normal unit vectors of M; the *support function* $p\colon \mathscr{D} \to \mathbf{R}$ be defined by

$$(3.25)\qquad p(d) = \langle v_3(M(d)), v(d)\rangle .$$

From (IV.2.6), we get

$$(3.26)\qquad {}^*p = \varepsilon_2 p ;$$

from Lemma IV.2.2,

$$(3.27)\qquad {}^*p{}^*H = pH ,$$

i.e., pH and p^2 are invariants of the couple (M, S). Further, let $\sigma(d) \geqq 0$ be defined by

$$(3.28)\qquad \sigma(d)^2 = |v(d)|^2 - p(d)^2 ;$$

$\sigma(d)$ is, of course, the length of the orthogonal projection of $v(d)$ into the tangent plane of M at $M(d)$.

Theorem 3.4. *Let $M\colon \mathscr{D} \to E^3$, $\mathscr{D} \subset \mathbf{R}^2$ a bounded domain, be a surface and $S \in E^3$ a fixed point. Let the couple (M, S) satisfy:* (i) *there are functions $F, P, Q\colon \mathscr{D} \to \mathbf{R}$ such that*

$$(3.29)\qquad Kp^2P + Hp(Q - P) - Q = \sigma^2 F , \qquad Kp^2P^2 + 2HpPQ + Q^2 > 0 ;$$

(ii) $\sigma = 0$, *on ∂M. Then M is a part of a sphere centered at S.*

Proof. Write

$$(3.30)\qquad v = xv_1 + yv_2 + pv_3 ;$$

from (3.24) and $\mathrm{d}S = 0$,

$$(3.31)\qquad \mathrm{d}x - y\omega_1{}^2 - p\omega_1{}^3 + \omega^1 = 0 , \quad \mathrm{d}y + x\omega_1{}^2 - p\omega_2{}^3 + \omega^2 = 0 ,$$
$$\mathrm{d}p + x\omega_1{}^3 + y\omega_2{}^3 = 0 .$$

Let the coordinates in $\mathscr{D}$ and the frames of M be chosen in such a way that (1.4) are valid. From (3.31),

$$(3.32)\qquad \frac{\partial x}{\partial u} + \frac{1}{s}\frac{\partial r}{\partial v}\, y = (pa - 1)\, r , \quad \frac{\partial x}{\partial v} - \frac{1}{r}\frac{\partial s}{\partial u}\, y = pbs ,$$
$$\frac{\partial y}{\partial u} - \frac{1}{s}\frac{\partial r}{\partial v}\, x = pbr , \quad \frac{\partial y}{\partial v} + \frac{1}{r}\frac{\partial s}{\partial u}\, x = (pc - 1)\, s .$$

From $(3.32_{2,3})$,

$$(3.33)\qquad r\frac{\partial x}{\partial v} - s\frac{\partial y}{\partial u} = -\frac{\partial r}{\partial v}\, x + \frac{\partial s}{\partial u}\, y .$$

Multiplying $(3.32_{1,2,4})$ by $(Pcp + Q)\, s$, $-2Pbpr$, $(Pap + Q)\, r$ resp. and adding them together, we get

$$(3.34)\quad (Pcp + Q)\, s \frac{\partial x}{\partial u} - 2Pbpr \frac{\partial x}{\partial v} + (Pap + Q)\, r \frac{\partial y}{\partial v}$$
$$+ (Pap + Q) \frac{\partial s}{\partial u} x + \left\{(Pcp + Q) \frac{\partial r}{\partial v} + 2Pbp \frac{\partial s}{\partial u}\right\} y$$
$$= 2\{Kp^2P + Hp(Q - P) - Q\}\, rs.$$

Now, $\sigma^2 = x^2 + y^2$, and the right-hand side of (3.34) may be written, because of (3.29_1), as $2rsxF \cdot x + 2rsyF \cdot y$. Thus (3.34) takes the form

$$(3.35)\quad (Pcp + Q)\, s \frac{\partial x}{\partial u} - 2Pbpr \frac{\partial x}{\partial r} + (Pap + Q)\, r \frac{\partial y}{\partial v} = (.)\, x + (.)\, y.$$

The system (3.33) + (3.35) has the form (III.3.1), the associated form (III.3.2) being

$$(3.36)\quad \Phi = (Pap + Q)\, r^2\mu^2 + 2Pbprs\mu\nu + (Pcp + Q)\, s^2\nu^2.$$

The discriminant of Φ is equal to

$$(3.37)\quad \Delta = (Kp^2P^2 + 2HpPQ + Q^2)\, r^2s^2;$$

(3.29_2) implies $\Delta > 0$. From (ii), $x = y = 0$ on ∂M. Thus $x = y = 0$ on M, and we are done. QED.

For $P = F = 0$, $Q = 1$ and $F = 0$, $P = Q = 1$ resp., we get two known consequences, which we present here in an even generalized form (the generalization removing the classical supposition $K > 0$).

Corollary 3.1 (Hp-theorem). *Let $M: \mathscr{D} \to E^3$, $\mathscr{D} \subset \mathbf{R}^2$ a bounded domain, be a surface and $S \in E^3$ a fixed point. Let the couple (M, S) satisfy:* (i) *on M,*

$$(3.38)\quad Hp = 1;$$

(ii) *$\sigma = 0$ on ∂M. Then M is a part of a sphere with the center at S.*

Corollary 3.2 (Kp-theorem). *Let $M: \mathscr{D} \to E^3$, $\mathscr{D} \subset \mathbf{R}^2$ a bounded domain, be a surface and $S \in E^3$ a fixed point. Let the couple (M, S) satisfy:* (i) *on M,*

$$(3.39)\quad Kp^2 = 1, \qquad Hp + 1 > 0;$$

(ii) *$\sigma = 0$ on ∂M. Then M is a part of a sphere with the center at S.*

4. Weingarten surfaces — maximum principle

To show the possibility of proving certain results by means of the maximum principle, we are going to produce various generalizations of the H- and K-theorems.

Theorem 4.1. *Let $M: \mathscr{D} \to E^3$ be a surface satisfying:* (i) *$K \geq 0$ on M;* (ii) *M has an orthogonal net of lines of curvature;* (iii) *v_1 and v_2 being the unit tangent vector fields* *of the lines of curvature and S a real valued function on M with*

$$|S| \leqq 4\sqrt{2} - 5, \tag{4.1}$$

we have

$$v_1 v_1 H - v_2 v_2 H + S[v_1, v_2]\, H = 0; \tag{4.2}$$

(iv) *∂M consists of umbilical points. Then M is a part of a sphere or a plane resp.*

Proof. On M, consider the function

$$f = 2(H^2 - K) = \frac{1}{2}(a - c)^2 + 2b^2; \tag{4.3}$$

of course, $f \geqq 0$ on M, and $f = 0$ exactly in the umbilical points. The covariant derivatives of f with respect to the frames $\{m, v_1, v_2, v_3\}$ being defined by (IV.3.15) and (IV.3.17), it is easy to calculate that

$$f_1 = (a - c)(\alpha - \gamma) + 4b\beta, \quad f_2 = (a - c)(\beta - \delta) + 4b\gamma; \tag{4.4}$$

$$\begin{aligned} f_{11} &= (c^2 - ac + 4b^2)K + (\alpha - \gamma)^2 + 4\beta^2 + (a - c)(A - C) + 4bB, \\ f_{12} &= 2b(a + c)K + (\alpha - \gamma)(\beta - \delta) + 4\beta\gamma + (a - c)(B - D) + 4bC, \\ f_{22} &= (a^2 - ac + 4b^2)K + (\beta - \delta)^2 + 4\gamma^2 + (a - c)(C - E) + 4bD. \end{aligned} \tag{4.5}$$

The vectors v_1, v_2 being principal, we have to suppose

$$b = 0; \tag{4.6}$$

of course, we could make the above calculations taking already in regard this simplification, but the formulas (4.4) and (4.5) will be very usefull in the future. From (4.6) and (IV.2.22) and (IV.2.24), we get (2.35) and

$$\begin{aligned} (a - c)\, v_1 v_1 a &= 3\beta^2 + (a - c)A, \quad (a - c)\, v_1 v_2 a = (2\gamma - \alpha)\beta + (a - c)B, \\ (a - c)\, v_2 v_1 a &= 3\beta\gamma + (a - c)B, \\ (a - c)\, v_2 v_2 a &= (2\gamma - \alpha)\gamma + (a - c)(C + aK), \\ (a - c)\, v_1 v_1 c &= (\delta - 2\beta)\beta + (a - c)(C + cK), \\ (a - c)\, v_1 v_2 c &= -3\beta\gamma + (a - c)D, \\ (a - c)\, v_2 v_1 c &= (\delta - 2\beta)\gamma + (a - c)D, \\ (a - c)\, v_2 v_2 c &= -3\gamma^2 + (a - c)E. \end{aligned} \tag{4.7}$$

Thus we have (2.11) and

$$\begin{aligned} (a - c)\, v_1 v_1 H &= \frac{1}{2}(\beta + \delta)\beta + \frac{1}{2}(a - c)(A + C + cK), \\ (a - c)\, v_1 v_2 H &= -\frac{1}{2}(\alpha + \gamma)\beta + \frac{1}{2}(a - c)(B + D), \\ (a - c)\, v_2 v_1 H &= \frac{1}{2}(\beta + \delta)\gamma + \frac{1}{2}(a - c)(B + D), \\ (a - c)\, v_2 v_2 H &= -\frac{1}{2}(\alpha + \gamma)\gamma + \frac{1}{2}(a - c)(C + E + aK). \end{aligned} \tag{4.8}$$

The equation (4.2) implies

$$(4.9) \qquad \beta^2 + \gamma^2 + \alpha\gamma + \beta\delta - (a-c)^2 K + (a-c)(A-E) - (S(\alpha\beta + 2\beta\gamma + \gamma\delta) = 0.$$

The elimination of A, C, E from (4.9) and $(4.5_{1,3})$ yields

$$(4.10) \qquad f_{11} + f_{22} - 4Kf = \left(\alpha - \frac{3}{2}\gamma + \frac{1}{2}S\beta\right)^2 + \left(\delta - \frac{3}{2}\beta + \frac{1}{2}S\gamma\right)^2 + \frac{1}{4}\{(7-S^2)\beta^2 + 20S\beta\gamma + (7-S^2)\gamma^2\}.$$

The last term of the right-hand side is non-negative, for each β and γ, because of (4.1). Our theorem follows then immediately from Theorem 1.2. QED.

Let us add the following remark. Let M satisfy (i), (ii) and (iv) of Theorem 4.1. On M, consider the second order operators of the form

$$(4.11) \qquad \mathscr{P} = S_1 v_1 v_1 + S_2 v_2 v_2 + S_3 v_1 v_2 + S_4 v_2 v_1,$$

$S_1, \ldots, S_4 \colon \mathscr{D} \to \mathbf{R}$ being functions and v_1, v_2 the unit tangent vector fields of the lines of curvature. Now, our problem is to determine the class of operators $\mathscr{P}$ (4.11) with the following property: The surface M satisfying $\mathscr{P}H = 0$ (and knowing nothing more about it), we are able to prove, by means of the maximum principle using the function $f = 2(H^2 - K)$, that it is a part of a sphere. It is not difficult to see that the just defined class of operators is given by (iii) of Theorem 4.1. Indeed, $\mathscr{P}H = 0$ implies

$$(4.12) \qquad (\beta+\delta)(\beta S_1 + \gamma S_4) - (\alpha+\gamma)(\beta S_3 + \gamma S_2) + (a-c)K(cS_1 + aS_2) + (a-c)\{(A+C)S_1 + (B+D)(S_3+S_4) + (C+E)S_2\} = 0.$$

To get a linear second order equation for f (4.3), we should be able to eliminate $A, \ldots, E$ from (4.5) and (4.12). Thus the matrix

$$(4.13) \qquad \begin{Vmatrix} 1 & 0 & -1 & 0 & 0 \\ 0 & 0 & 0 & -1 & 0 \\ 0 & 0 & 1 & 0 & -1 \\ S_1 & S_3 + S_4 & S_1 + S_2 & S_3 + S_4 & S_2 \end{Vmatrix}$$

should have its rang equal to 3. This condition implies $S_1 + S_2 = S_3 + S_4 = 0$, and we are done.

We are now going to pursue the matter taken in this remark. Again, consider a surface M satisfying conditions (i), (ii), (iv) of Theorem 4.1. Let $\varphi \colon \mathbf{R}^2 \to \mathbf{R}$ be a function (it is sufficient that f be defined on a set containing $a(M) \times c(M)$) and the induced function $\varphi \colon \mathscr{D} \to \mathbf{R}$ given by $\varphi(a, c)$. From

$$(4.14) \qquad \mathrm{d}\varphi = \frac{\partial \varphi}{\partial a}(\alpha\omega^1 + \beta\omega^2) + \frac{\partial \varphi}{\partial c}(\gamma\omega^1 + \delta\omega^2),$$

we obtain

$$(4.15)\qquad v_1\varphi = \frac{\partial\varphi}{\partial a}\,\alpha + \frac{\partial\varphi}{\partial c}\,\gamma, \qquad v_2\varphi = \frac{\partial\varphi}{\partial a}\,\beta + \frac{\partial\varphi}{\partial c}\,\delta.$$

From the expression for $\mathrm{d}(v_1\varphi)$ and $\mathrm{d}(v_2\varphi)$, we get

$$(4.16)\qquad (a-c)\,v_1v_1\varphi = (a-c)\left(\alpha^2\frac{\partial^2\varphi}{\partial a^2} + 2\alpha\gamma\frac{\partial^2\varphi}{\partial a\,\partial c} + \gamma^2\frac{\partial^2\varphi}{\partial c^2} + cK\frac{\partial\varphi}{\partial c} + A\frac{\partial\varphi}{\partial a} + C\frac{\partial\varphi}{\partial c}\right) + \left(3\beta\frac{\partial\varphi}{\partial a} + \delta\frac{\partial\varphi}{\partial c} - 2\beta\frac{\partial\varphi}{\partial c}\right)\beta,$$

$$(a-c)\,v_2v_1\varphi = (a-c)\left(\alpha\beta\frac{\partial^2\varphi}{\partial a^2} + \alpha\delta\frac{\partial^2\varphi}{\partial a\,\partial c} + \beta\gamma\frac{\partial^2\varphi}{\partial a\,\partial c} + \gamma\delta\frac{\partial^2\varphi}{\partial c^2} + B\frac{\partial\varphi}{\partial a} + D\frac{\partial\varphi}{\partial c}\right) + \left(3\beta\frac{\partial\varphi}{\partial a} + \delta\frac{\partial\varphi}{\partial c} - 2\beta\frac{\partial\varphi}{\partial c}\right)\gamma,$$

$$(a-c)\,v_1v_2\varphi = (a-c)\left(\alpha\beta\frac{\partial^2\varphi}{\partial a^2} + \beta\gamma\frac{\partial^2\varphi}{\partial a\,\partial c} + \alpha\delta\frac{\partial^2\varphi}{\partial a\,\partial c} + \gamma\delta\frac{\partial^2\varphi}{\partial c^2} + B\frac{\partial\varphi}{\partial a} + D\frac{\partial\varphi}{\partial c}\right) + \left(2\gamma\frac{\partial\varphi}{\partial a} - \alpha\frac{\partial\varphi}{\partial a} - 3\gamma\frac{\partial\varphi}{\partial c}\right)\beta,$$

$$(a-c)\,v_2v_2\varphi = (a-c)\left(\beta^2\frac{\partial^2\varphi}{\partial a^2} + 2\beta\delta\frac{\partial^2\varphi}{\partial a\,\partial c} + \delta^2\frac{\partial^2\varphi}{\partial c^2} + aK\frac{\partial\varphi}{\partial a} + C\frac{\partial\varphi}{\partial a} + E\frac{\partial\varphi}{\partial c}\right) + \left(2\gamma\frac{\partial\varphi}{\partial a} - \alpha\frac{\partial\varphi}{\partial a} - 3\gamma\frac{\partial\varphi}{\partial c}\right)\gamma.$$

Let us consider the second order operator

$$(4.17)\qquad \mathscr{P}_R = v_1v_1 - v_2v_2 + R[v_1, v_2]$$

for functions on M; here, $R\colon M \to \mathbf{R}$ is a given function. It is easy to see that

$$(4.18)\qquad \frac{\partial\varphi}{\partial a}f_{11} + \frac{\partial\varphi}{\partial c}f_{22} - 2\left(\frac{\partial\varphi}{\partial a} + \frac{\partial\varphi}{\partial c}\right)Kf = (a-c)\,\mathscr{P}_R\varphi - \Phi_1 + \Phi_2$$

with

$$\Phi_1 = (a-c)\left\{(\alpha^2-\beta^2)\frac{\partial^2\varphi}{\partial a^2} + 2(\alpha\gamma-\beta\delta)\frac{\partial^2\varphi}{\partial a\,\partial c} + (\gamma^2-\delta^2)\frac{\partial^2\varphi}{\partial c^2}\right\},$$

$$\Phi_2 = (\alpha^2+\beta^2+3\gamma^2-3\alpha\gamma+R\alpha\beta+R\beta\gamma)\frac{\partial\varphi}{\partial a} + (3\beta^2+\gamma^2+\delta^2-3\beta\delta+R\gamma\delta+R\beta\gamma)\frac{\partial\varphi}{\partial c}.$$

This equation is the basis of our further considerations. Of course, we have just proved the following very general (the notation being obvious)

Theorem 4.2. *Let $M: \mathscr{D} \to E^3$ be a surface satisfying:* (i) $K \geqq 0$; (ii) *M has an orthogonal net of lines of curvature;* (iii) *there is a function $\varphi(a, c)$ such that*

$$\frac{\partial \varphi}{\partial a} > 0, \quad \frac{\partial \varphi}{\partial c} = 0, \quad (a - c)\, \mathscr{P}_R \varphi \geqq 0, \quad \Phi_2 \geqq \Phi_1; \tag{4.19}$$

(iv) *∂M consists of umbilical points. Then M is a part of a sphere or a plane resp.*

Let us consider some corollaries of Theorem 4.2. First of all, take

$$\varphi(a, c) = T_1 a + T_2 c; \qquad T_1, T_2 \in \mathbf{R}. \tag{4.20}$$

Then

$$\begin{aligned} \Phi_2 - \Phi_1 = {} & T_1 \left(\alpha + \frac{1}{2} R\beta - \frac{3}{2}\gamma\right)^2 + T_2 \left(\delta - \frac{3}{2}\beta + \frac{1}{2} R\gamma\right)^2 \\ & + \frac{1}{4} \{(4 - R^2)\, T_1 + 3T_2\}\, \beta^2 + \frac{5}{2} (T_1 + T_2)\, R\beta\gamma \\ & + \frac{1}{4} \{3T_1 + (4 - R^2)\, T_2\}\, \gamma^2, \end{aligned} \tag{4.21}$$

and we get the following

Corollary 4.1. *Let $M: \mathscr{D} \to E^3$ be a surface satisfying the conditions* (i), (ii), (iv) *of* Theorem 4.2 *and* (iii$_1$): *there are constants $T_1 > 0$, $T_2 > 0$ and a function $R: \mathscr{D} \to \mathbf{R}$ such that*

$$(a - c)\, \mathscr{P}_R(T_1 a + T_2 c) \geqq 0, \tag{4.22}$$

$$\begin{aligned} & R^2 \leqq \min\,(4 + 3T_1^{-1} T_2,\ 4 + 3T_1 T_2^{-2}), \\ & T_1 T_2 R^4 - (28T_1^2 + 58T_1 T_2 + 28T_2^2)\, R^2 + 12T_1^2 + 25T_1 T_2 + 12T_2^2 \geqq 0. \end{aligned} \tag{4.23}$$

Then M is a part of a sphere or a plane resp.

Notice that, for $T_1 = T_2 = 1$, we have recovered Theorem 4.1 even in a generalized form, the equality (4.2) being replaced by

$$(a - c)\, (v_1 v_1 H - v_2 v_2 H + S[v_1, v_2]\, H) \geqq 0. \tag{4.24}$$

Let us take

$$\varphi(a, c) = ac = K, \tag{4.25}$$

and suppose $K \neq 0$. Then (4.18) takes the form

$$c f_{11} + a f_{22} - 4HKf = (a - c)\, \mathscr{P}_R K + \Phi_2 - \Phi_1 \tag{4.26}$$

with

$$\begin{aligned} \Phi_2 - \Phi_1 = {} & c \left\{\alpha + \frac{1}{2}\left(R\beta - \frac{2a + c}{c}\gamma\right)\right\}^2 + a \left\{\delta + \frac{1}{2}\left(R\gamma - \frac{a + 2c}{a}\beta\right)\right\}^2 \\ & + \frac{1}{4a} (11a^2 - 4c^2 - acR^2)\, \beta^2 + \frac{5}{2} (a + c)\, R\beta\gamma \\ & + \frac{1}{4c} (11c^2 - 4a^2 - acR^2)\, \gamma^2. \end{aligned}$$

From this, we get

Corollary 4.2. *Let $M: \mathscr{D} \to E^3$ be a surface satisfying the conditions* (ii), (iv) *of* Theorem 4.2 *and* (i_2) $K > 0$; (iii_2) *there is a function $R: \mathscr{D} \to \mathbf{R}$ such that*

$$(4.27) \qquad (a - c)\, \mathscr{P}_R K \geqq 0,$$

$$(4.28) \qquad R^2 \leqq \min\left(11\frac{a}{c} - 4\frac{c}{a},\, 11\frac{c}{a} - 4\frac{a}{c}\right),$$

$$a^2c^2R^4 - 2ac\,(16a^2 + 25ac + 16c^2)\, R^2 - 44a^4 + 137a^2c^2 - 44c^4 \geqq 0.$$

Then M is a part of a sphere.

Up to now, we have considered surfaces possessing an orthogonal net of lines of curvature. Let us present two results for general surfaces, the first one being a generalization of the K-theorem and the second one of the H-theorem.

As usually, the covariant derivatives of the Gauss' curvature with respect to a given field of frames are given by

$$(4.29) \qquad \mathrm{d}K = K_1\omega^1 + K_2\omega^2;$$

$$\mathrm{d}K_1 - K_2\omega_1{}^2 = K_{11}\omega^1 + K_{12}\omega^2, \qquad \mathrm{d}K_2 + K_1\omega_1{}^2 = K_{12}\omega^1 + K_{22}\omega^2.$$

It is just a matter of patience to see that $(a - c)(K_{11} - K_{22}) + 4bK_{12}$ is an invariant up to a sign induced by the orientation of the field of normal vectors.

Theorem 4.3. *Let $M: \mathscr{D} \to E^3$ be a surface satisfying:* (i) $K > 0$ *on* M; (ii) *the unit normal vectors of M being chosen in such a way that $H > 0$ on M, we have*

$$(4.30) \qquad (a - c)(K_{11} - K_{22}) + 4bK_{12} \geqq 0$$

on M; (iii) *for each point $m \in M$, $k_1 \leqq k_2$ being the principal curvatures at m, we have* $k_2 \leqq 3k_1$; (iv) *∂M consists of umbilical points. Then M is a part of a sphere.*

Proof. It is easy to see that

$$(4.31) \qquad K_1 = a\gamma - 2b\beta + c\alpha, \qquad K_2 = a\delta - 2b\gamma + c\beta;$$

$$(4.32) \qquad K_{11} = (ac - 2b^2)\, K + 2\alpha\gamma - 2\beta^2 + cA - 2bB + aC,$$

$$K_{12} = -b(a + c)\, K + \alpha\delta - \beta\gamma + cB - 2bC + aD,$$

$$K_{22} = (ac - 2b^2)\, K + 2\beta\delta - 2\gamma^2 + cC - 2bD + aE.$$

Multiplying the equations (4.5) and (4.32) successively by c, $-2b$, a, $c - a$, $-4b$, $a - c$ and adding them together, we get

$$(4.33) \qquad cf_{11} - 2bf_{12} + af_{22} - 4HKf = (a - c)(K_{11} - K_{22}) + 4bK_{12} + T$$

with

$$T = a(3\beta^2 + \delta^2 + 2\gamma^2 - 2\alpha\gamma) - 2b(\alpha + \gamma)(\beta + \delta)$$
$$+ c(\alpha^2 + 3\gamma^2 + 2\beta^2 - 2\beta\delta).$$

We have to prove $T \geqq 0$. The term T is an invariant of our surface (the proof of this being a computational matter only); thus it is sufficient to prove $T \geqq 0$ in a generic point $m \in M$ using a convenable field of moving frames around m. As always, let us use a field of frames with $b(m) = 0$. Then, at m,

$$(4.34)\quad T = \frac{1}{c}(c\alpha - a\gamma)^2 + \frac{1}{a}(a\delta - c\beta)^2 + 2H\left\{\frac{1}{a}(3a - c)\beta^2 + \frac{1}{c}(3c - a)\gamma^2\right\};$$

$a(m) > 0$ and $c(m) > 0$ being the principal curvatures, we are done. QED.

The classical K-theorem is a direct consequence of (4.33). At $m \in M$ with $b(m) = 0$, we have

$$(4.35)\quad K_1 = a\gamma + c\alpha = 0, \quad K_2 = a\delta + c\beta = 0$$

as a consequence of $K = \text{const}$; thus

$$(4.36)\quad T(m) = (3a^2 + 2ac + 3c^2)(a\sigma^2 + c\varrho^2) \geqq 0,$$

$\varrho, \sigma \in \mathbf{R}$ being given by $\alpha = \varrho\alpha$, $\gamma = -\varrho c$, $\beta = \sigma a$, $\delta = -\sigma c$.

Theorem 4.4. *Let $M: \mathscr{D} \to E^3$ be a surface satisfying:* (i) $K > 0$ *on* M; (ii) *we have*

$$(4.37)\quad (a - c)(H_{11} - H_{22}) + 4bH_{12} + H_1^2 + H_2^2 \geqq 0$$

on M, the left-hand side of (4.37) *being an invariant of M;* (iii) ∂M *consists of umbilical points. Then M is a part of a sphere.*

Proof. From (2.11), we get

$$(4.38)\quad 2H_{11} = cK + A + C, \quad 2H_{12} = B + D, \quad 2H_{22} = aK + C + E.$$

Multiplying the equations (4.5) and (4.38) successively by $1, 0, 1, c - a, -4b, a - c$ and adding them together, we get

$$(4.39)\quad f_{11} + f_{22} - 4Kf = 2\{(a - c)(H_{11} - H_{22}) + 4bH_{12} + H_1^2 + H_2^2\} + \frac{1}{2}\{(3\gamma - \alpha)^2 + (3\beta - \delta)^2\}.$$

Our result follows then from (4.37) and Theorem 1.2. QED.

5. Harmonic mappings

The purpose of the present paragraph is to give another generalization of the H-theorem.

Let $M \subset E^3$ be a surface and let $v_3(m)$ be a unit normal vector at $m \in M$; let us choose a fixed field of unit normal vectors along M. Choose a point $s \in E^3$, and let $S^2 \subset E^3$ be the unit sphere centered at s. We naturally get the *spherical mapping* $f: M \to S^2$, $m \mapsto v_3(m)$; we are going to study this mapping quite systematically. Instead of S^2, we are going to take into account an arbitrary surface M^* and $f: M \to M^*$ is to be an arbitrary mapping.

Let $M, M^* \subset E^3$ be surfaces; let $\{m, v_1, v_2, v_3\}$, $\{m^*, v_1^*, v_2^*, v_3^*\}$ be associated fields of orthonormal tangent frames satisfying

$$(5.1)\qquad \begin{aligned} dm &= \omega^1 v_1 + \omega^2 v_2, & dm^* &= \Omega^1 v_1^* + \Omega^2 v_2^*,\\ dv_1 &= \omega_1{}^2 v_2 + \omega_1{}^3 v_3, & dv_1^* &= \Omega_1{}^2 v_2^* + \Omega_1{}^3 v_3^*,\\ dv_2 &= -\omega_1{}^2 v_1 + \omega_2{}^3 v_3, & dv_2^* &= -\Omega_1{}^2 v_1^* + \Omega_2{}^3 v_3^*,\\ dv_3 &= -\omega_1{}^3 v_1 - \omega_2{}^3 v_2, & dv_3^* &= -\Omega_1{}^3 v_1^* - \Omega_2{}^3 v_2^*. \end{aligned}$$

Suppose that the surfaces M and M^* are *oriented* in the sense that we admit just the following changes of tangent frames:

$$(5.2)\qquad \begin{aligned} &\tilde v_1 = \cos\alpha \cdot v_1 - \sin\alpha \cdot v_2,\quad \tilde v_2 = \sin\alpha\cdot v_1 + \cos\alpha\cdot v_2,\quad \tilde v_3 = \varepsilon v_3;\quad \varepsilon^2 = 1;\\ &\tilde v_1^* = \cos\alpha^*\cdot v_1^* - \sin\alpha^*\cdot v_2^*,\quad \tilde v_2^* = \sin\alpha^*\cdot v_1^* + \cos\alpha^*\cdot v_2^*,\\ &\tilde v_3^* = \varepsilon^* v_3^*;\quad \varepsilon^{*2} = 1. \end{aligned}$$

Be given a mapping $f\colon M \to M^*$; on M, we get the induced forms

$$(5.3)\qquad \tau^i = f^*(\Omega^i),\quad \tau_i{}^j = f^*(\Omega_i{}^j);$$

of course

$$(5.4)\qquad \tau^i(v) = \Omega^i(f_* v),\quad \tau_i{}^j(v) = \Omega_i{}^j(f_* v)$$

for each tangent vector $v \in T_m(M)$. The forms τ^i, $\tau_i{}^j$ satisfy the equations

$$(5.5)\qquad \tau^3 = 0,\quad \tau_i{}^j + \tau_j{}^i = 0,\quad d\tau^i = \sum_{j=1}^{3} \omega^j \wedge \omega_j{}^i,\quad d\tau_i{}^j = \sum_{k=1}^{3} \tau_i{}^k \wedge \tau_k{}^j,$$

this following from the general properties of the mapping f^* introduced in Chap. I.

Let us write

$$(5.6)\qquad \tau^1 = a_1\omega^1 + a_2\omega^2,\quad \tau^2 = a_3\omega^1 + a_4\omega^2,$$

and let us suppose that $f\colon M \to M^*$ is *orientation preserving* in the sense that we assume

$$(5.7)\qquad \mu := a_1 a_4 - a_2 a_3 \geqq 0$$

at each point $m \in M$. Let $v = xv_1 + yv_2 \in T_m(M)$, and let

$$(5.8)\qquad f_* v_1 = \tilde a_1 v_1^* + \tilde a_2 v_2^*,\quad f_* v_2 = \tilde a_3 v_1^* + \tilde a_4 v_2^*.$$

From (5.4),

$$\begin{aligned} (a_1\omega^1 + a_2\omega^2)(xv_1 + yv_2) &= \Omega^1\{x(\tilde a_1 v_1^* + \tilde a_2 v_2^*) + y(\tilde a_3 v_1^* + \tilde a_4 v_2^*)\},\\ (a_3\omega^1 + a_4\omega^2)(xv_1 + yv_2) &= \Omega^2\{x(\tilde a_1 v_1^* + \tilde a_2 v_2^*) + y(\tilde a_3 v_1^* + \tilde a_4 v_2^*)\}, \end{aligned}$$

i.e.,

$$a_1 x + a_2 y = \tilde a_1 x + \tilde a_3 y,\quad a_3 x + a_4 y = \tilde a_2 x + \tilde a_4 y$$

for each $x, y \in \mathbf{R}$. Thus (1.8) become

$$(5.9)\qquad f_* v_1 = a_1 v_1^* + a_3 v_2^*,\quad f_* v_2 = a_2 v_1^* + a_4 v_2^*.$$

From (5.6), we get

$$(5.10)\quad \begin{aligned} &(\mathrm{d}a_1 - a_2\omega_1{}^2 - a_3\tau_1{}^2) \wedge \omega^1 + (\mathrm{d}a_2 + a_1\omega_1{}^2 - a_4\tau_1{}^2) \wedge \omega^2 = 0,\\ &(\mathrm{d}a_3 - a_4\omega_1{}^2 + a_1\tau_1{}^2) \wedge \omega^1 + (\mathrm{d}a_4 + a_3\omega_1{}^2 + a_2\tau_1{}^2) \wedge \omega^2 = 0 \end{aligned}$$

and the existence of functions $b_1, \ldots, b_6$ such that

$$(5.11)\quad \begin{aligned} &\mathrm{d}a_1 - a_2\omega_1{}^2 - a_3\tau_1{}^2 = b_1\omega^1 + b_2\omega^2,\\ &\mathrm{d}a_2 + a_1\omega_1{}^2 - a_4\tau_1{}^2 = b_2\omega^1 + b_3\omega^2,\\ &\mathrm{d}a_3 - a_4\omega_1{}^2 + a_1\tau_1{}^2 = b_4\omega^1 + b_5\omega^2,\\ &\mathrm{d}a_4 + a_3\omega_1{}^2 + a_2\tau_1{}^2 = b_5\omega^1 + b_6\omega^2. \end{aligned}$$

Further, K and K^* being the Gauss curvatures of M and M^* resp.,

$$(5.12)\quad \begin{aligned} &(\mathrm{d}b_1 - 2b_2\omega_1{}^2 - b_4\tau_1{}^2) \wedge \omega^1 + \{\mathrm{d}b_2 + (b_1 - b_3)\,\omega_1{}^2 - b_5\tau_1{}^2\} \wedge \omega^2\\ &= (a_2K + a_3\mu K^*)\,\omega^1 \wedge \omega^2,\\ &\{\mathrm{d}b_2 + (b_1 - b_3)\,\omega_1{}^2 - b_5\tau_1{}^2\} \wedge \omega^1 + \{\mathrm{d}b_3 + 2b_2\omega_1{}^2 - b_6\tau_1{}^2\} \wedge \omega^2\\ &= (-a_1K + a_4\mu K^*)\,\omega^1 \wedge \omega^2,\\ &(\mathrm{d}b_4 - 2b_5\omega_1{}^2 + b_1\tau_1{}^2) \wedge \omega^1 + \{\mathrm{d}b_5 + (b_4 - b_6)\,\omega_1{}^2 + b_2\tau_1{}^2\} \wedge \omega^2\\ &= (a_4K - a_1\mu K^*)\,\omega^1 \wedge \omega^2,\\ &\{\mathrm{d}b_5 + (b_4 - b_6)\,\omega_1{}^2 + b_2\tau_1{}^2\} \wedge \omega^1 + (\mathrm{d}b_6 + 2b_5\omega_1{}^2 + b_3\tau_1{}^2) \wedge \omega^2\\ &= (-a_3K - a_2\mu K^*)\,\omega^1 \wedge \omega^2. \end{aligned}$$

Here, of course, $\mathrm{d}\tau_1{}^2 = -K^*\tau^1 \wedge \tau^2 = -K^*\mu\omega^1 \wedge \omega^2$. Thus we get the existence of functions $c_1, \ldots, c_8$ such that

$$(5.13)\quad \begin{aligned} &\mathrm{d}b_1 - 2b_2\omega_1{}^2 - b_4\tau_1{}^2 = c_1\omega^1 + c_2\omega^2,\\ &\mathrm{d}b_2 + (b_1 - b_3)\,\omega_1{}^2 - b_5\tau_1{}^2 = (c_2 + a_2K + a_3\mu K^*)\,\omega^1\\ &\qquad\qquad + (c_3 + a_1K - a_4\mu K^*)\,\omega^2,\\ &\mathrm{d}b_3 + 2b_2\omega_1{}^2 - b_6\tau_1{}^2 = c_3\omega^1 + c_4\omega^2,\\ &\mathrm{d}b_4 - 2b_5\omega_1{}^2 + b_1\tau_1{}^2 = c_5\omega^1 + c_6\omega^2,\\ &\mathrm{d}b_5 + (b_4 - b_6)\,\omega_1{}^2 + b_2\tau_1{}^2 = (c_6 + a_4K - a_1\mu K^*)\,\omega^1\\ &\qquad\qquad + (c_7 + a_3K + a_2\mu K^*)\,\omega^2,\\ &\mathrm{d}b_6 + 2b_5\omega_1{}^2 + b_3\tau_1{}^2 = c_7\omega^1 + c_8\omega^2. \end{aligned}$$

Changing the frames of both surfaces by means of (5.2), we get

$$(5.14)\quad \begin{aligned} &\omega^1 = \cos\alpha\cdot\tilde\omega^1 + \sin\alpha\cdot\tilde\omega^2, &&\omega^2 = -\sin\alpha\cdot\tilde\omega^1 + \cos\alpha\cdot\tilde\omega^2;\\ &\tau^1 = \cos\alpha^*\cdot\tilde\tau^1 + \sin\alpha^*\cdot\tilde\tau^2, &&\tau^2 = -\sin\alpha^*\cdot\tilde\tau^1 + \cos\alpha^*\cdot\tilde\tau^2. \end{aligned}$$

Substituting into (5.6) and using analogous equations for $\tilde\tau^1$, $\tilde\tau^2$, we get

$$\begin{aligned} &\cos\alpha^*\cdot(\tilde a_1\tilde\omega^1 + \tilde a_2\tilde\omega^2) + \sin\alpha^*\cdot(\tilde a_3\tilde\omega^1 + \tilde a_4\tilde\omega^2)\\ &= a_1(\cos\alpha\cdot\tilde\omega^1 + \sin\alpha\cdot\tilde\omega^2) + a_2(-\sin\alpha\cdot\tilde\omega^1 + \cos\alpha\cdot\tilde\omega^2),\\ &\quad -\sin\alpha^*\cdot(\tilde a_1\tilde\omega^1 + \tilde a_2\tilde\omega^2) + \cos\alpha^*\cdot(\tilde a_3\tilde\omega^1 + \tilde a_4\tilde\omega^2)\\ &= a_3(\cos\alpha\cdot\tilde\omega^1 + \sin\alpha\cdot\tilde\omega^2) + a_4(-\sin\alpha\cdot\tilde\omega^1 + \cos\alpha\cdot\tilde\omega^2), \end{aligned}$$

i.e.,

$$(5.15)\quad \begin{aligned} a_1 &= \cos\alpha\cdot\cos\alpha^*\cdot\tilde a_1 + \sin\alpha\cdot\cos\alpha^*\cdot\tilde a_2 + \cos\alpha\cdot\sin\alpha^*\cdot\tilde a_3 \\ &\quad + \sin\alpha\cdot\sin\alpha^*\cdot\tilde a_4, \\ a_2 &= -\sin\alpha\cdot\cos\alpha^*\cdot\tilde a_1 + \cos\alpha\cdot\cos\alpha^*\cdot\tilde a_2 - \sin\alpha\cdot\sin\alpha^*\cdot\tilde a_3 \\ &\quad + \cos\alpha\cdot\sin\alpha^*\cdot\tilde a_4, \\ a_3 &= -\cos\alpha\cdot\sin\alpha^*\cdot\tilde a_1 - \sin\alpha\cdot\sin\alpha^*\cdot\tilde a_2 + \cos\alpha\cdot\cos\alpha^*\cdot\tilde a_3 \\ &\quad + \sin\alpha\cdot\cos\alpha^*\cdot\tilde a_4, \\ a_4 &= \sin\alpha\cdot\sin\alpha^*\cdot\tilde a_1 - \cos\alpha\cdot\sin\alpha^*\cdot\tilde a_2 - \sin\alpha\cdot\cos\alpha^*\cdot\tilde a_3 \\ &\quad + \cos\alpha\cdot\cos\alpha^*\cdot\tilde a_4. \end{aligned}$$

Lemma 5.1. *Let*

$$(5.16)\quad \begin{aligned} &I_1 = a_1^2 + a_2^2 + a_3^2 + a_4^2, \\ &I_2 = (a_1 - a_4)^2 + (a_2 + a_3)^2, \qquad I_3 = (a_1 + a_4)^2 + (a_2 - a_3)^2. \end{aligned}$$

Then

$$(5.17)\quad \tilde I_1 = I_1, \quad \tilde I_2 = I_2, \quad \tilde I_3 = I_3, \quad \tilde\mu = \mu.$$

Of course, $I_2 = I_1 - 2\mu$, $I_3 = I_1 - 2\mu$.

Recall the transformation formulas — see (IV.2.10) —

$$(5.18)\quad \tilde\omega_1{}^2 = \omega_1{}^2 - \mathrm{d}\alpha, \qquad \tilde\tau_1{}^2 = \tau_1{}^2 - \mathrm{d}\alpha^*.$$

Then (5.11) and (5.14) + (5.15) imply

$$(5.19)\quad \begin{aligned} b_1 &= \cos^2\alpha\cos\alpha^*\cdot\tilde b_1 + 2\sin\alpha\cos\alpha\cos\alpha^*\cdot\tilde b_2 + \sin^2\alpha\cos\alpha^*\cdot\tilde b_3 \\ &\quad + \cos^2\alpha\sin\alpha^*\cdot\tilde b_4 + 2\sin\alpha\cos\alpha\sin\alpha^*\cdot\tilde b_5 + \sin^2\alpha\sin\alpha^*\cdot\tilde b_6, \\ b_2 &= -\sin\alpha\cos\alpha\cos\alpha^*\cdot\tilde b_1 + (\cos^2\alpha - \sin^2\alpha)\cos\alpha^*\cdot\tilde b_2 \\ &\quad + \sin\alpha\cos\alpha\cos\alpha^*\cdot\tilde b_3 - \sin\alpha\cos\alpha\sin\alpha^*\cdot\tilde b_4 \\ &\quad + (\cos^2\alpha - \sin^2\alpha)\sin\alpha^*\cdot\tilde b_5 + \sin\alpha\cos\alpha\sin\alpha^*\cdot\tilde b_6, \\ b_3 &= \sin^2\alpha\cos\alpha^*\cdot\tilde b_1 - 2\sin\alpha\cos\alpha\cos\alpha^*\cdot\tilde b_2 + \cos^2\alpha\cos\alpha^*\cdot\tilde b_3 \\ &\quad + \sin^2\alpha\sin\alpha^*\cdot\tilde b_4 - 2\sin\alpha\cos\alpha\sin\alpha^*\cdot\tilde b_5 + \cos^2\alpha\sin\alpha^*\cdot\tilde b_6, \\ b_4 &= -\cos^2\alpha\sin\alpha^*\cdot\tilde b_1 - 2\sin\alpha\cos\alpha\sin\alpha^*\cdot\tilde b_2 - \sin^2\alpha\sin\alpha^*\cdot\tilde b_3 \\ &\quad + \cos^2\alpha\cos\alpha^*\cdot\tilde b_4 + 2\sin\alpha\cos\alpha\cos\alpha^*\cdot\tilde b_5 + \sin^2\alpha\cos\alpha^*\cdot\tilde b_6, \\ b_5 &= \sin\alpha\cos\alpha\sin\alpha^*\cdot\tilde b_1 + (\sin^2\alpha - \cos^2\alpha)\sin\alpha^*\cdot\tilde b_2 \\ &\quad - \sin\alpha\cos\alpha\sin\alpha^*\cdot\tilde b_3 - \sin\alpha\cos\alpha\cos\alpha^*\cdot\tilde b_4 \\ &\quad + (\cos^2\alpha - \sin^2\alpha)\cos\alpha^*\cdot\tilde b_5 + \sin\alpha\cos\alpha\cos\alpha^*\cdot\tilde b_6, \\ b_6 &= -\sin^2\alpha\sin\alpha^*\cdot\tilde b_1 + 2\sin\alpha\cos\alpha\sin\alpha^*\cdot\tilde b_2 - \cos^2\alpha\sin\alpha^*\cdot\tilde b_3 \\ &\quad + \sin^2\alpha\cos\alpha^*\cdot\tilde b_4 - 2\sin\alpha\cos\alpha\cos\alpha^*\cdot\tilde b_5 + \cos^2\alpha\cos\alpha^*\cdot\tilde b_6. \end{aligned}$$

From these,

$$(5.20)\quad \begin{aligned} b_1 + b_3 &= \cos \alpha^* \cdot (\tilde{b}_1 + \tilde{b}_3) + \sin \alpha^* \cdot (\tilde{b}_4 + \tilde{b}_6), \\ b_4 + b_6 &= -\sin \alpha^* \cdot (\tilde{b}_1 + \tilde{b}_3) + \cos \alpha^* \cdot (\tilde{b}_4 + \tilde{b}_6). \end{aligned}$$

From (5.13),

$$(5.21)\quad \begin{aligned} \mathrm{d}(b_1 + b_3) - (b_4 + b_6)\, \tau_1{}^2 &= (c_1 + c_3)\, \omega^1 + (c_2 + c_4)\, \omega^2, \\ \mathrm{d}(b_4 + b_6) + (b_1 + b_3)\, \tau_1{}^2 &= (c_5 + c_7)\, \omega^1 + (c_6 + c_8)\, \omega^2, \end{aligned}$$

and we get

$$(5.22)\quad \begin{aligned} c_1 + c_3 &= \cos\alpha \cos\alpha^* \cdot (\tilde{c}_1 + \tilde{c}_3) + \sin\alpha \cos\alpha^* \cdot (\tilde{c}_2 + \tilde{c}_4) \\ &\quad + \cos\alpha \sin\alpha^* \cdot (\tilde{c}_5 + \tilde{c}_7) + \sin\alpha \sin\alpha^* \cdot (\tilde{c}_6 + \tilde{c}_8), \\ c_2 + c_4 &= -\sin\alpha \cos\alpha^* \cdot (\tilde{c}_1 + \tilde{c}_3) + \cos\alpha \cos\alpha^* \cdot (\tilde{c}_2 + \tilde{c}_4) \\ &\quad - \sin\alpha \sin\alpha^* \cdot (\tilde{c}_5 + \tilde{c}_7) + \cos\alpha \sin\alpha^* \cdot (\tilde{c}_6 + \tilde{c}_8), \\ c_5 + c_7 &= -\cos\alpha \sin\alpha^* \cdot (\tilde{c}_1 + \tilde{c}_3) - \sin\alpha \sin\alpha^* \cdot (\tilde{c}_2 + \tilde{c}_4) \\ &\quad + \cos\alpha \cos\alpha^* \cdot (\tilde{c}_5 + \tilde{c}_7) + \sin\alpha \cos\alpha^* \cdot (\tilde{c}_6 + \tilde{c}_8), \\ c_6 + c_8 &= \sin\alpha \sin\alpha^* \cdot (\tilde{c}_1 + \tilde{c}_3) - \cos\alpha \sin\alpha^* \cdot (\tilde{c}_2 + \tilde{c}_4) \\ &\quad - \sin\alpha \cos\alpha^* \cdot (\tilde{c}_5 + \tilde{c}_7) + \cos\alpha \cos\alpha^* \cdot (\tilde{c}_6 + \tilde{c}_8). \end{aligned}$$

Finally, for $\varepsilon = \pm 1$,

$$(5.23)\quad \begin{aligned} c_1 + c_3 + \varepsilon(c_6 + c_8) &= (\cos\alpha \cos\alpha^* + \varepsilon \sin\alpha \sin\alpha^*)\, \{\tilde{c}_1 + \tilde{c}_3 + \varepsilon(\tilde{c}_6 + \tilde{c}_8)\} \\ &\quad + (\sin\alpha \cos\alpha^* - \varepsilon \cos\alpha \sin\alpha^*)\, \{\tilde{c}_2 + \tilde{c}_4 - \varepsilon(\tilde{c}_5 + \tilde{c}_7)\} \\ c_2 + c_4 - \varepsilon(c_5 + c_7) &= (-\sin\alpha \cos\alpha^* + \varepsilon \cos\alpha \sin\alpha^*)\, \{\tilde{c}_1 + \tilde{c}_3 + \varepsilon(\tilde{c}_6 + \tilde{c}_8)\} \\ &\quad + (\cos\alpha \cos\alpha^* + \varepsilon \sin\alpha \sin\alpha^*)\, \{\tilde{c}_2 + \tilde{c}_4 - \varepsilon(\tilde{c}_5 + \tilde{c}_7)\}. \end{aligned}$$

After these preliminary computations, we are in the position to present the following definitions. The terms *harmonic, totally geodesic* and *tension field* are well established in the literature, the term *ε-weakly harmonic* is new.

Definition 5.1. *The mapping* $f\colon M \to M^*$ *is called* (i) constant, (ii) conformal, (iii) harmonic, (iv) ε-weakly harmonic ($\varepsilon = \pm 1$) *if* (i) $I_1 = 0$, (ii) $I_2 = 0$, (iii)

$$(5.24)\quad b_1 + b_3 = b_4 + b_6 = 0,$$

(iv)

$$(5.25)\quad c_1 + c_3 + \varepsilon(c_6 + c_8) = c_2 + c_4 - \varepsilon(c_5 + c_7) = 0$$

resp.

Notice that a harmonic mapping f is ε-weakly harmonic.

Definition 5.2. *Be given a mapping* $f\colon M \to M^*$. *The mapping* $t\colon M \to T(M^*)$, $t(m) \in T_{f(m)}(M^*)$, *given by*

$$(5.26)\quad t = (b_1 + b_3)\, v_1{}^* + (b_4 + b_6)\, v_2{}^*$$

is called the tension field *of* f.

Both the definitions are all right from the formal point of view; nevertheless, we have to explain them geometrically.

It is easy to see that f is constant if and only if $f(M)$ is just a point. Further,

$$\text{(5.27)} \qquad \mathrm{d}s_*^2 = (a_1^2 + a_3^2)(\omega^1)^2 + 2(a_1a_2 + a_3a_4)\,\omega^1\omega^2 + (a_2^2 + a_4^2)(\omega^2)^2;$$

thus the term conformal is to be meant in the normal sense.

Now, let $f: M \to M^*$ be a (general) given mapping, let $m_0 \in M$ be a fixed point. Let $L: E^3 \to E^3$ be an affine mapping such that

$$\text{(5.28)} \qquad Lm_0 = f(m_0), \qquad L|_{T_{m_0}(M)} = (f_*)_{m_0}, \qquad Lv_3 = \pm v_3^*.$$

Such a mapping is given by

$$\text{(5.29)} \qquad Lm_0 = m_0^*, \quad Lv_1 = a_1v_1^* + a_3v_2^*, \quad Lv_2 = a_2v_1^* + a_4v_2^*, \quad Lv_3 = \pm v_3^*.$$

Let $m = m(s)$ be a curve on M through m_0, $v = x(s)\,v_1 + y(s)\,v_2$ its tangent vector at $m(s)$, s its arc and let $m^* = m^*(s)$ be its image in f. Then it is easy to show that, at m_0,

$$\text{(5.30)} \qquad \frac{\mathrm{d}^2m^*}{\mathrm{d}s^2} - L\,\frac{\mathrm{d}^2m}{\mathrm{d}s^2} = \frac{\mathscr{L}(v)}{I(v)} + \frac{\mathscr{L}_0(v)}{I(v)}\,v_3,$$

where

$$\text{(5.31)} \qquad \mathscr{L}: T_m(M) \to T_{f(m)}(M^*)$$

is the quadratic mapping given by

$$\text{(5.32)} \qquad \mathscr{L}(xv_1 + yv_2) = (b_1x^2 + 2b_2xy + b_3y^2)\,v_1^* + (b_4x^2 + 2b_5xy + b_6y^2)\,v_2^*$$

and

$$\text{(5.33)} \qquad \mathscr{L}_0: T_m(M) \to \mathbf{R}$$

is the quadratic mapping given by

$$\text{(5.34)} \qquad \mathscr{L}_0(v) = \tau^1(v)\,\tau_1^3(v) + \tau^2(v)\,\varepsilon_2^3(v) \mp \omega^1(v)\,\omega_1^3(v) \mp \omega^2(v)\,\omega_2^3(v).$$

We shall be mainly interested in the mapping (5.31).

On each $T_m(M)$, consider the mapping

$$\text{(5.35)} \qquad *: T_m(M) \to T_m(M), \qquad *\,(xv_1 + yv_2) = yv_1 - xv_2;$$

it is easy to see that this definition of the $*$-operator is a good one. Of course, this $*$-operator is the adjoint operator to the $*$-operator on 1-forms; namely, we have

$$\text{(5.36)} \qquad *\omega(v) = \omega(*v)$$

for each 1-form ω and each vector v at $m \in M$. A similar $*$-operator is to be considered on each $T_{m^*}(M^*)$. The following proposition shows one of the possible geometric definitions of the tension field.

Proposition 5.1. *Let $v \in T_m(M)$ be any tangent unit vector at m. Then*

$$\text{(5.37)} \qquad t(m) = \mathscr{L}(v) + \mathscr{L}(*v).$$

The geometrical interpretation of the conditions (5.25) is explained in the following

Proposition 5.2. *For a vector w of E^3 with the origin at $m^* \in M^*$, denote by w^T its tangential part, i.e., its orthogonal projection into $T_{m^*}(M^*)$. Then: The mapping $f\colon M \to M^*$ is ε-weakly harmonic if and only if, for each $m \in M$ and each $v \in T_m(M)$,*

$$(5.38) \qquad [\{f_*(*v)\}\, t + \varepsilon * \{(f_* v)\, t\}]^T = 0,$$

t being the tension field of f.

Proof. Let $v = xv_1 + yv_2$. It is easy to see that

$$(5.39) \qquad (f_* v)\, t = \{(c_1 + c_3)\, x + (c_2 + c_4)\, y\}\, v_1^* + \{(c_5 + c_7)\, x + (c_6 + c_8)\, y\}\, v_2^* \\ + \{(b_1 + b_3)\, \tau_1^3(v) + (b_4 + b_6)\, \tau_2^3(v)\}\, v_3^*;$$

from this, we see that the left-hand side of (5.38) is equal to

$$(5.40) \qquad \{c_1 + c_3 + \varepsilon(c_6 + c_8)\}\, (yv_1^* - \varepsilon x v_2^*) - \{c_2 + c_4 - \varepsilon(c_5 + c_7)\}\, (xv_1^* + \varepsilon y v_2^*),$$

and our proposition follows. QED.

The following lemma follows immediately from (5.19) and (5.22).

Lemma 5.2. *Let*

$$(5.41) \qquad \begin{aligned} L_1 &= (b_1 - b_5)\, (b_3 + b_5) - (b_2 - b_6)\, (b_2 + b_4), \\ L_2 &= (b_1 + b_5)\, (b_3 - b_5) - (b_2 + b_6)\, (b_2 - b_4), \\ L_3 &= (c_1 + c_3)\, (c_6 + c_8) - (c_2 + c_4)\, (c_5 + c_7). \end{aligned}$$

Then

$$(5.42) \qquad L_1 = \tilde{L}_1, \quad L_2 = \tilde{L}_2, \quad L_3 = \tilde{L}_3.$$

We are going to present the geometrical signification of the invariants (5.39). To do this, consider the bilinear mapping

$$(5.43) \qquad \mathscr{L}\colon T_m(M) \times T_m(M) \to T_{f(m)}(M^*)$$

associated to (5.31); (5.43) is given by

$$(5.44) \qquad \mathscr{L}(xv_1 + yv_2, x'v_1 + y'v_2) = (b_1 xx' + b_2 xy' + b_2 x'y + b_3 yy')\, v_1^* \\ + (b_4 xx' + b_5 xy' + b_5 x'y + b_6 yy')\, v_2^*.$$

Proposition 5.3. *Let $v \in T_m(M)$ be an arbitrary unit vector. Then*

$$(5.45) \qquad \begin{aligned} L_1 &= -\langle \mathscr{L}(v, v) + *\mathscr{L}(v, *v), \mathscr{L}(v, *v) + *\mathscr{L}(*v, *v)\rangle, \\ L_2 &= -\langle \mathscr{L}(v, v) - *\mathscr{L}(v, *v), *\mathscr{L}(*v, *v) - \mathscr{L}(v, *v)\rangle, \\ L_3 &= -\varepsilon (f_* v)\, t, \{f_*(*v)\}\, t\rangle. \end{aligned}$$

Proof. Because of the invariance of L_1, L_2, L_3, we may assume $v = v_1$, and the proof follows immediately from (5.44) and (5.39). QED.

Finally, we get the following

Lemma 5.3. *The expressions*

$$(5.46) \quad J_1 = (b_1 + b_3)^2 + (b_4 + b_6)^2, \quad J_2 = b_1^2 + 2b_2^2 + b_3^2 + b_4^2 + 2b_5^2 + b_6^2$$

are invariants of f.

Our next result is auxiliary.

Proposition 5.4. *The condition*

(5.47) f *is harmonic*

and/or the condition

(5.48) *for each* $m \in M$ *there is* $\dim \mathscr{L}(T_m(M)) \leqq 1$ *and there exists a vector* $0 \neq v \in T_m(M)$ *such that* $\mathscr{L}(v) = 0$

implies

$$(5.49) \quad L_1 \leqq 0, \quad L_2 \leqq 0.$$

Proof. The condition (5.47) is equivalent to (5.24). Hence

$$L_1 = -(b_1 - b_5)^2 - (b_2 + b_4)^2 \leqq 0, \quad L_2 = -(b_1 + b_2)^2 - (b_2 - b_4)^2 \leqq 0.$$

From (5.48), we get the existence of functions ϱ, σ, B_1, B_2, B_3 such that

$$b_1 = \varrho B_1, \quad b_2 = \varrho B_2, \quad b_3 = \varrho B_3, \quad b_4 = \sigma B_1, \quad b_5 = \sigma B_2, \quad b_6 = \sigma B_3,$$

i.e.,

$$\mathscr{L}(xv_1 + yv_2) = (B_1x^2 + 2B_2xy + B_3y^2)(\varrho v_1^* + \sigma v_2^*).$$

Further,

$$L_1 = L_2 = (\varrho^2 + \sigma^2)(B_1B_3 - B_2^2),$$

and our lemma follows. QED.

After these preliminary remarks, we may prove a series of results. In all theorems, we have in our mind a given mapping $f: M \to M^*$.

Theorem 5.1. *Suppose:* (i) $L_1 \leqq 0$ *on* M; (ii) $K > 0$ *on* M; (iii) $K^* \geqq 0$ *on* $f(M) \subset M^*$; (iv) $I_2 = 0$ *on* ∂M. *Then* f *is conformal.*

Proof. Consider, on M, the 1-form

$$(5.50) \quad \varphi_1 = \{(a_1 - a_4)(b_2 + b_4) - (a_2 + a_3)(b_1 - b_5)\}\,\omega^1 + \{(a_1 - a_4)(b_3 + b_5) - (a_2 + a_3)(b_2 - b_6)\}\,\omega^2.$$

It is easy to prove that this form is invariant and

$$(5.51) \quad d\varphi_1 = \{2L_1 - I_2(K + \mu K^*)\}\,\omega^1 \wedge \omega^2.$$

The result follows from the integral formula $\int_{\partial M} \varphi_1 = \int_M d\varphi_1$. QED.

Theorem 5.2. *Suppose:* (i) $L_2 \leqq 0$ *on* M; (ii) $K > 0$ *on* M; (iii) $K^* \leqq 0$ *on* $f(M) \subset M^*$; (iv) $J_2 = 0$ *on* ∂M. *Then* f *is a constant mapping.*

Proof. Consider the invariant 1-form

$$\begin{aligned}(5.52)\qquad \varphi_2 = {} & \{(a_1 + a_4)(b_2 - b_4) - (a_2 - a_3)(b_1 + b_5)\}\,\omega^1 \\ & + \{(a_1 + a_4)(b_3 - b_5) - (a_2 - a_3)(b_2 + b_6)\}\,\omega^2\end{aligned}$$

with

$$(5.53)\qquad \mathrm{d}\varphi_2 = \{2L_2 + I_3(\mu K^* - K)\}\,\omega^1 \wedge \omega^2.$$

Thus $I_3 = 0$, i.e., $a_1 + a_4 = a_2 - a_3 = 0$. Hence $\mu = -a_1^2 - a_2^2$; from $\mu \geqq 0$, we get $a_1 = a_2 = 0$. QED.

Theorem 5.3. *Suppose:* (i) f *is* (-1)*-weakly harmonic*; (ii) $K > 0$ *on* M; (iii) $K^* \geqq 0$ *on* $f(M) \subset M^*$; (iv) $I_2 = 0$ *on* ∂M. *Then* f *is conformal.*

Proof. We have

$$\begin{aligned}(5.54)\qquad \frac{1}{2}\,\mathrm{d}I_2 = {} & \{(a_1 - a_4)(b_1 - b_5) + (a_2 + a_3)(b_2 + b_4)\}\,\omega^1 \\ & + \{(a_1 - a_4)(b_2 - b_6) + (a_2 + a_3)(b_3 + b_5)\}\,\omega^2\end{aligned}$$

and the integral formula

$$\begin{aligned}(5.55)\qquad \frac{1}{2}\int_{\partial M} * \,\mathrm{d}I_2 = \int_M & \{(a_1 - a_4)(c_1 + c_3 - c_6 - c_8) + (a_2 + a_3)(c_2 + c_4 + c_5 + c_7) \\ & + (b_1 - b_5)^2 + (b_2 + b_4)^2 + (b_2 - b_6)^2 + (b_3 + b_5)^2 \\ & + I_2(K + \mu K^*)\}\,\omega^1 \wedge \omega^2.\end{aligned}$$

Our theorem follows. QED.

Weakening the condition (ii), we get

Theorem 5.4. *Suppose:* (i) f *is* (-1)*-weakly harmonic*; (ii) $K \geqq 0$ *on* M; (iii) $K^* \geqq 0$ *on* $f(M) \subset M^*$; (iv) $I_2 = 0$ *on* ∂M. *Then* f *is harmonic and we have*

$$(5.56)\qquad I_2(K + \mu K^*) = 0$$

at each point $m \in M$.

Proof. From (5.55),

$$b_1 - b_5 = b_2 + b_4 = b_2 - b_6 = b_3 + b_5 = 0$$

and f is harmonic. From (5.13),

$$\begin{aligned} c_1 - c_6 &= a_4 K - a_1 \mu K^*, & c_2 - c_7 &= -a_2 K - a_3 \mu K^*, \\ c_2 - c_7 &= a_3 K + a_2 \mu K^*, & c_3 - c_8 &= -a_1 K + a_4 \mu K^*, \\ c_2 + c_5 &= -a_2 K - a_3 \mu K^*, & c_3 + c_6 &= -a_4 K + a_1 \mu K^*, \\ c_3 + c_6 &= -a_1 K + a_4 \mu K^*, & c_4 + c_7 &= -a_3 K - a_2 \mu K^*. \end{aligned}$$

Eliminating $c_1, \ldots, c_8$ from these equations and $(5.25)_{\varepsilon=-1}$, we get

$$(a_1 - a_4)(K + \mu K^*) = (a_2 + a_3)(K + \mu K^*) = 0,$$

i.e., (5.56). QED.

Theorem 5.5. *Suppose:* (i) f *is* $(+1)$*-weakly harmonic*; (ii) $K > 0$ *on* M; (iii) $K^* \leqq 0$ *on* $f(M) \subset M^*$; (iv) $J_2 = 0$ *on* ∂M. *Then* f *is a constant mapping.*

Proof. We have

$$(5.57) \quad \frac{1}{2} \mathrm{d}I_3 = \{(a_1 + a_4)(b_1 + b_5) + (a_2 - a_3)(b_2 - b_4)\} \omega^1 + \{(a_1 + a_4)(b_2 + b_6) + (a_2 - a_3)(b_3 - b_5)\} \omega^2$$

and the integral formula

$$(5.58) \quad \frac{1}{2} \int_{\partial M} * \mathrm{d}I_3 = \int_M \{(a_1 + a_4)(c_1 + c_3 + c_6 + c_8) + (a_2 - a_3)(c_2 + c_4 - c_5 - c_7) + (b_1 + b_5)^2 + (b_2 - b_4)^2 + (b_2 + b_6)^2 + (b_3 - b_5)^2 + I_3(K - \mu K^*)\} \omega^1 \wedge \omega^2.$$

Thus $I_3 = 0$, i.e., $\mu = -a_1^2 - a_2^2$. From $\mu \geqq 0$, $a_1 = a_2 = 0$. QED.

Theorem 5.6. *Suppose:* (i) f *is* $(+1)$*-weakly harmonic*; (ii) $K \geqq 0$ *on* M; (iii) $K^* \leqq 0$ *on* $f(M) \subset M^*$; (iv) $J_2 = 0$ *on* ∂M. *Then* f *is harmonic and we have*

$$(5.59) \quad I_3(K - \mu K^*) = 0$$

at each point $m \in M$.

Proof. From (5.58),

$$b_1 + b_5 = b_2 - b_4 = b_2 + b_6 = b_3 - b_5 = 0,$$

and f is harmonic. Further,

$$c_1 + c_6 = -a_4 K + a_1 \mu K^*, \qquad c_2 + c_7 = -a_2 K - a_3 \mu K^*,$$
$$c_2 + c_7 = -a_3 K - a_2 \mu K^*, \qquad c_3 + c_8 = -a_1 K + a_4 \mu K^*,$$
$$c_2 - c_5 = -a_2 K - a_3 \mu K^*, \qquad c_3 - c_5 = a_4 K - a_1 \mu K^*,$$
$$c_3 - c_6 = -a_1 K + a_4 \mu K^*, \qquad c_4 - c_7 = a_3 K + a_2 \mu K^*.$$

The elimination of $c_1, \ldots, c_8$ from these equations and from $(5.25)_{\varepsilon=1}$ implies

$$(a_1 + a_4)(K - \mu K^*) = (a_2 - a_3)(K - \mu K^*) = 0,$$

i.e., (5.59). QED.

Theorem 5.7. *Suppose:* (i) M *is compact*; (ii) $J_1 = \text{const} \neq 0$ *on* M; (iii) $K^* > 0$ *or* $K^* < 0$ *on* $f(M) \subset M^*$. *Then* $\dim f_*(T_m(M)) \leqq 1$ *for each* $m \in M$.

Proof. From $J_1 = \text{const} \neq 0$, we get

$$(b_1 + b_3)(c_1 + c_3) + (b_4 + b_6)(c_5 + c_7) = 0,$$
$$(b_1 + b_3)(c_2 + c_4) + (b_4 + b_6)(c_6 + c_8) = 0$$

and the existence of ϱ, σ such that

$$c_1 + c_3 = \varrho(b_4 + b_6), \qquad c_5 + c_7 = -\varrho(b_1 + b_3),$$
$$c_2 + c_4 = \sigma(b_4 + b_6), \qquad c_6 + c_8 = -\sigma(b_1 + b_3),$$

i.e., $L_3 = 0$.

From (5.21),

$$\{\mathrm{d}(c_1 + c_3) - (c_2 + c_4)\,\omega_1{}^2 - (c_5 + c_7)\,\tau_1{}^2\} \wedge \omega^1$$
$$+ \{\mathrm{d}(c_2 + c_4) + (c_1 + c_3)\,\omega_1{}^2 - (c_6 + c_8)\,\tau_1{}^2\} \wedge \omega^2 = (b_4 + b_6)\,\mu K^*\omega^1 \wedge \omega^2,$$
$$\{\mathrm{d}(c_5 + c_7) - (c_6 + c_8)\,\omega_1{}^2 + (c_1 + c_3)\,\tau_1{}^2\} \wedge \omega^1$$
$$+ \{\mathrm{d}(c_6 + c_8) + (c_5 + c_7)\,\omega_1{}^2 + (c_2 + c_4)\,\tau_1{}^2\} \wedge \omega^2 = -(b_1 + b_3)\,\mu K^*\omega^1 \wedge \omega^2,$$

and we get the existence of functions $e_1, \ldots, e_6$ such that

$$\mathrm{d}(c_1 + c_3) - (c_2 + c_4)\,\omega_1{}^2 - (c_5 + c_7)\,\tau_1{}^2 = e_1\omega^1 + (e_2 - b_4\mu K^*)\,\omega^2,$$
$$\mathrm{d}(c_2 + c_4) + (c_1 + c_3)\,\omega_1{}^2 - (c_6 + c_8)\,\tau_1{}^2 = (e_2 + b_6\mu K^*)\,\omega^1 + e_3\omega^2,$$
$$\mathrm{d}(c_5 + c_7) - (c_6 + c_8)\,\omega_1{}^2 + (c_1 + c_3)\,\tau_1{}^2 = e_4\omega^1 + (e_5 + b_1\mu K^*)\,\omega^2,$$
$$\mathrm{d}(c_6 + c_8) + (c_5 + c_7)\,\omega_1{}^2 + (c_2 + c_4)\,\tau_1{}^2 = (e_5 - b_3\mu K^*)\,\omega^1 + e_6\omega^2.$$

It may be easily shown that the 1-form

$$(5.60) \qquad \varphi_3 = \{(b_1 + b_3)(c_5 + c_7) - (b_4 + b_6)(c_1 + c_3)\}\,\omega^1 + \{(b_1 + b_3)(c_6 + c_8) - (b_4 + b_6)(c_2 + c_4)\}\,\omega^2$$

is invariant, and we get the integral formula

$$(5.61) \qquad \int_{\partial M} \varphi_3 = \int_M (2L_3 - J_1\mu K^*)\,\omega^1 \wedge \omega^2.$$

From $L_3 = 0$, $\mu = 0$ on M. QED.

Now, let us return to a spherical mapping of a surface. In E^3, let us choose a fixed point S_0, let S^2 be the unit sphere centered at S_0, and let $f\colon M \to S^2$, $m^* = f(m) = S_0 - v_3(m)$, be the spherical mapping. Then

$$(5.62) \qquad \begin{aligned} \mathrm{d}m^* &= \omega_1{}^3 v_1 + \omega_2{}^3 v_2, \\ \mathrm{d}v_1 &= \qquad\quad \omega_1{}^2 v_2 + \omega_1{}^3 v_3, \\ \mathrm{d}v_2 &= -\omega_1{}^2 v_1 \qquad\quad + \omega_2{}^3 v_3, \\ \mathrm{d}v_3 &= -\omega_1{}^3 v_1 - \omega_2{}^3 v_2, \end{aligned}$$

i.e.,

$$(5.63) \qquad v_1^* = v_1, \quad v_2^* = v_2, \quad v_3^* = v_3;$$

$$(5.64) \qquad \begin{aligned} &\tau^1 = \omega_1{}^3 = a\omega^1 + b\omega^2, \quad \tau^2 = \omega_2{}^3 = b\omega^1 + c\omega^2, \quad \tau_1{}^2 = \omega_1{}^2; \\ &a_1 = a, \quad a_2 = a_3 = b, \quad a_4 = c; \\ &K^* = 1, \quad \mu = ac - b^2 = K; \\ &I_1 = a^2 + 2b^2 + c^2 = 2(2H^2 - K), \quad I_2 = (a - c)^2 + 4b^2 = 4(H^2 - K), \\ &I_3 = (a + c)^2 = 4H^2. \end{aligned}$$

Further, comparing the equations (5.11), (5.13) with (IV.2.22) and (IV.2.24) resp., we get

$$(5.65) \quad b_1 = \alpha, \quad b_2 = \beta, \quad b_3 = \gamma, \quad b_4 = \beta, \quad b_5 = \gamma, \quad b_6 = \delta;$$

$$c_1 = A, \quad c_2 = B - bK, \quad c_3 = C + cK, \quad c_4 = D + bK,$$

$$c_5 = B + bK, \quad c_6 = C + aK, \quad c_7 = D - bK, \quad c_8 = E;$$

$$t = (\alpha + \gamma)\, v_1 + (\beta + \delta)\, v_2;$$

$$\mathscr{L}(xv_1 + yv_2, x'v_1 + y'v_2)$$
$$= (\alpha xx' + \beta xy' + \beta x'y + \gamma yy')\, v_1 + (\beta xx' + \gamma xy' + \gamma x'y + \delta yy')\, v_2;$$

$$J_1 = (\alpha + \gamma)^2 + (\beta + \delta)^2, \quad J_2 = \alpha^2 + 3\beta^2 + 3\gamma^2 + \delta^2.$$

From these, we get the following (local)

Proposition 5.5. *The spherical mapping $f\colon M \to S^2$ of a surface is* (i) *constant,* (ii) *conformal,* (iii) *harmonic if and only if the surface M is a part of a* (i) *plane,* (ii) *sphere,* (iii) *surface with constant mean curvature resp.*

The H-theorem for ovaloids follows easily. Indeed, let M be a surface with constant mean curvature and $K > 0$, and let ∂M consist of umbilical points. According to the last proposition, its spherical mapping $f\colon M \to S^2$ is harmonic and we have $I_2 = 0$ on ∂M. The mapping f being harmonic, we get $L_1 \leqq 0$ from Proposition 5.4, and f is conformal according to Theorem 5.1. Now, go back to Proposition 5.5, (ii). We might apply Theorem 5.3 as well: f being harmonic, it is, of course, (-1)-weakly harmonic.

The reader is invited to work out other consequences of the theorems of this paragraph.

6. Differential equations on surfaces

Up to now, we have been concerned with the *existence* of (roughly speaking) ovaloids satisfying certain conditions imposed on the curvatures, namely, we have proved that ovaloids satisfying these conditions reduce to spheres. The purpose of the present paragraph is to establish a *non-existence* theorem.

Suppose that M is a surface in E^3; let $f\colon M \to \mathbf{R}$ be a function. Let us restrict our study to a domain of M covered by a field of orthonormal tangent frames. To f, we get by the standard prolongation procedure the covariant derivatives f_i; f_{ij}; $M, \ldots, Q$; $R, \ldots, V$ of f with respect to the given field of frames by means of the formulas

$$(6.1) \quad \mathrm{d}f = f_1\omega^1 + f_2\omega^2,$$

$$(6.2) \quad (\mathrm{d}f_1 - f_2\omega_1^{\;2}) \wedge \omega^1 + (\mathrm{d}f_2 + f_1\omega_1^{\;2}) \wedge \omega^2 = 0,$$

$$(6.3) \quad \mathrm{d}f_1 - f_2\omega_1^{\;2} = f_{11}\omega^1 + f_{12}\omega^2, \quad \mathrm{d}f_2 + f_1\omega_1^{\;2} = f_{12}\omega^1 + f_{22}\omega^2,$$

$$(6.4) \quad (\mathrm{d}f_{11} - 2f_{12}\omega_1^{\;2}) \wedge \omega^1 + \{\mathrm{d}f_{12} + (f_{11} - f_{22})\,\omega_1^{\;2}\} \wedge \omega^2 = Kf_2\omega^1 \wedge \omega^2,$$
$$\{\mathrm{d}f_{12} + (f_{11} - f_{22})\,\omega_1^{\;2}\} \wedge \omega^1 + (\mathrm{d}f_{22} + 2f_{12}\omega_1^{\;2}) \wedge \omega^2 = -Kf_1\omega^1 \wedge \omega^2,$$

$$\text{(6.5)}\qquad \mathrm{d}f_{11} - 2f_{12}\omega_1{}^2 = M\omega^1 + \left(N - \frac{1}{2}Kf_2\right)\omega^2,$$

$$\mathrm{d}f_{12} + (f_{11} - f_{22})\,\omega_1{}^2 = \left(N + \frac{1}{2}Kf_2\right)\omega^1 + \left(P + \frac{1}{2}Kf_1\right)\omega^2,$$

$$\mathrm{d}f_{22} + 2f_{12}\omega_1{}^2 = \left(P - \frac{1}{2}Kf_1\right)\omega^1 + Q\omega^2,$$

$$\text{(6.6)}\qquad \left\{\mathrm{d}M - \left(3N + \frac{1}{2}Kf_2\right)\omega_1{}^2\right\}\wedge\omega^1 + \left\{\mathrm{d}N + \left(M - 2P - \frac{1}{2}Kf_1\right)\omega_1{}^2\right\}\wedge\omega^2$$

$$= \frac{1}{2}\,(5Kf_{12} + K_1f_2)\,\omega^1\wedge\omega^2,$$

$$\left\{\mathrm{d}N + \left(M - 2P - \frac{1}{2}Kf_1\right)\omega_1{}^2\right\}\wedge\omega^1$$

$$+ \left\{\mathrm{d}P + \left(2N - Q + \frac{1}{2}Kf_2\right)\omega_1{}^2\right\}\wedge\omega^2$$

$$= \frac{1}{2}\,(3Kf_{22} - 3Kf_{11} + K_2f_2 - K_1f_1)\,\omega^1\wedge\omega^2,$$

$$\left\{\mathrm{d}P + \left(2N - Q + \frac{1}{2}Kf_2\right)\omega_1{}^2\right\}\wedge\omega^1 + \left\{\mathrm{d}Q + \left(3P + \frac{1}{2}Kf_1\right)\omega_1{}^2\right\}\wedge\omega^2$$

$$= -\frac{1}{2}\,(5Kf_{12} + K_2f_1)\,\omega^1\wedge\omega^2,$$

$$\text{(6.7)}\qquad \mathrm{d}M - \left(3N + \frac{1}{2}Kf_2\right)\omega_1{}^2 = R\omega^1 + \left(S - \frac{5}{4}Kf_{12} - \frac{1}{4}K_1f_2\right)\omega^2,$$

$$\mathrm{d}N + \left(M - 2P - \frac{1}{2}Kf_1\right)\omega_1{}^2 = \left(S + \frac{5}{4}Kf_{12} + \frac{1}{4}K_1f_2\right)\omega^1$$

$$+ \left(T + \frac{3}{2}Kf_{11} + \frac{1}{2}K_1f_1\right)\omega^2,$$

$$\mathrm{d}P + \left(2N - Q + \frac{1}{2}Kf_2\right)\omega_1{}^2 = \left(T + \frac{3}{2}Kf_{22} + \frac{1}{2}K_2f_2\right)\omega^1$$

$$+ \left(U + \frac{5}{4}Kf_{12} + \frac{1}{4}K_2f_1\right)\omega^2,$$

$$\mathrm{d}Q + \left(3P + \frac{1}{2}Kf_1\right)\omega_1{}^2 = \left(U - \frac{5}{4}Kf_{12} - \frac{1}{4}K_2f_1\right)\omega^1 + V\omega^2;$$

here, as usually, K_i are defined by $\mathrm{d}K = K_1\omega^1 + K_2\omega^2$. The Laplace operator on functions on M being again defined by

$$\text{(6.8)}\qquad \Delta f\cdot\omega^1\wedge\omega^2 = \mathrm{d}*\mathrm{d}f,$$

we have

$$\Delta f = f_{11} + f_{22}. \tag{6.9}$$

On M, consider the 1-form

$$\varphi(f) = \left\{-f_{12}M + (f_{11} - f_{22})\left(N + \frac{1}{2}Kf_2\right) + f_{12}\left(P - \frac{1}{2}Kf_1\right)\right\}\omega^1 + \left\{-f_{12}\left(N - \frac{1}{2}Kf_2\right) + (f_{11} - f_{22})\left(P + \frac{1}{2}Kf_1\right) + f_{12}Q\right\}\omega^2. \tag{6.10}$$

It is just the matter of mechanical computations to prove

Lemma 6.1. *The form $\varphi(f)$ is invariant on an inwardly oriented surface and we have*

$$d\varphi(f) = -[\Phi(f) + \{(f_{11} - f_{22})^2 + 4f_{12}^2\}K]\,\omega^1 \wedge \omega^2 \tag{6.11}$$

with

$$\Phi(f) = 2(N^2 + P^2 - MP - NQ) - \frac{1}{2}(f_1^2 + f_2^2)K^2 - (f_1M + f_2Q)K. \tag{6.12}$$

Now, we are in the position to prove

Theorem 6.1. *Let $g\colon (0, \infty) \to \langle 0, \infty)$, $g \not\equiv 0$, be a function. Then there is no ovaloid $M \subset E^3$ such that*

$$\Delta\left\{\int K^{-1}\frac{dg(K)}{dK}\,dK\right\} + 2g(K) = 0. \tag{6.13}$$

Especially, there are no ovaloids $M \subset E^3$ satisfying

$$\alpha\Delta K^{\alpha-1} + (\alpha - 1)K^\alpha = 0, \qquad 1 \neq \alpha \in \mathbf{R}. \tag{6.14}$$

Proof. Evidently, (6.14) is a consequence of (6.13) for $g(K) = K^\alpha$.
Let us study, on M, a general equation

$$\Delta f + 2g = 0, \tag{6.15}$$

i.e.,

$$f_{11} + f_{22} + 2g = 0. \tag{6.16}$$

Because of (6.5), the differential consequences of (6.16) are

$$M + P - \frac{1}{2}Kf_1 + 2g_1 = 0, \qquad N + Q - \frac{1}{2}Kf_2 + 2g_2 = 0 \tag{6.17}$$

and $\Phi(f)$ (6.12) turns out to be

$$\begin{aligned}\Phi &= 2N^2 + 2P^2 - \frac{1}{2}(f_1^2 + f_2^2)K^2 \\ &\quad + (2P + f_1K)\left(P - \frac{1}{2}Kf_1 + 2g_1\right) + (2N + f_2K)\left(N - \frac{1}{2}Kf_2 + 2g_2\right) \\ &= (2N + g_2)^2 + (2P + g_1)^2 - (g_2 - f_2K)^2 - (g_1 - f_1K)^2.\end{aligned} \tag{6.18}$$

Suppose

(6.19) $\mathrm{d}g = K\,\mathrm{d}f;$

we get

(6.20) $\mathrm{d}K \wedge \mathrm{d}f = 0$

as an immediate consequence. The suppositions $f = f(K)$ and (6.19) imply

(6.21) $\Phi \geqq 0,$

and we get

(6.22) $f_{11} - f_{22} = 0, \quad f_{12} = 0$

from the integral formula $\int_M \mathrm{d}\varphi = 0$. From (6.16) and (6.22), $f_{11} = f_{22} = -g$, i.e.,

(6.23) $\mathrm{d}f = f_1\omega^1 + f_2\omega^2, \quad \mathrm{d}f_1 - f_2{\omega_1}^2 = -g\omega^1, \quad \mathrm{d}f_2 + f_1{\omega_1}^2 = -g\omega^2.$

Now,

(6.24) $\mathrm{d} * \mathrm{d}f = \mathrm{d}(-f_2\omega^1 + f_1\omega^2) = -2g\omega^1 \wedge \omega^2;$

the supposition $g \geqq 0$ and the integral formula $\int_M \mathrm{d} * \mathrm{d}f = 0$ imply

(6.25) $g = 0.$

Because of $K > 0$, we get, from (6.19),

(6.26) $f = \text{const}.$

Thus there is only one couple of functions $g(K)$, $f(K) = \int K^{-1} \cdot \mathrm{d}g(K)/\mathrm{d}K \cdot \mathrm{d}K$ satisfying (6.15), this couple being given by (6.25) + (6.26). QED.

The just proved theorem shows the importance (and the beauty) of the study of the partial differential equations on (compact) surfaces. Because of the availability of methods introduced above, let us pursue this matter in the case of a sphere.

Thus, let $S^2 \subset E^3$ be the unit sphere centered at the origin 0 of E^3. A function $f: S^2 \to \mathbf{R}$ is called *linear* if $f(m) = \langle m, a \rangle$, m denoting the radius vector of the point $m \in S^2$ and a being a constant vector in E^3. Let us introduce the following operators on real-valued functions on S^2:

(6.27) $\mathscr{L}f = f_{11} + f_{22} + 2f, \quad \mathscr{M}f = f_{11}f_{22} - f_{12}^2 + f(f_{11} + f_{22} + f).$

Obviously, $\mathscr{L}f = \Delta f + 2f$ does not depend on the choice of the frames, and it is easy to see that the same is true for $\mathscr{M}f$; $\mathscr{M}f$ is the so-called *Weingarten operator*. We have

(6.28) $(\mathscr{L}f)^2 - 4\mathscr{M}f = (f_{11} - f_{22})^2 + 4f_{12}^2 \geqq 0.$

Theorem 6.2. *Let $\mathscr{D} \subset S^2$ be a domain, $\partial\mathscr{D}$ its boundary and $f: \mathscr{D} \to \mathbf{R}$ a function.* (i) *If*

(6.29) $\mathscr{L}f = 0 \quad in \quad \mathscr{D},$

$$(6.30)\qquad \mathscr{M}f = 0 \quad in \quad \partial\mathscr{D},$$

f is linear. (ii) *Let* $F\colon \mathbf{R} \to \mathbf{R}$ *be a function satisfying, for each* $t \in \mathbf{R}$,

$$(6.31)\qquad F(t) > \frac{\mathrm{d}F(t)}{\mathrm{d}t}\left(t - \frac{\mathrm{d}F(t)}{\mathrm{d}t}\right) \quad or \quad F(t) = 0 \quad resp.$$

If

$$(6.32)\qquad \mathscr{M}f = F(\mathscr{L}f) \quad in \quad \mathscr{D},$$

$$(\mathscr{L}f)^2 - 4F(\mathscr{L}f) = 0 \quad in \quad \partial\mathscr{D},$$

f is linear.

Proof. In the formulas (6.5) and (6.7), we have to set $K = 1$, $K_1 = K_2 = 0$. The formulas (6.11) and (6.12) take the form

$$(6.33)\qquad \varphi(f) = \left\{(f_{11} - f_{22})\left(N + \frac{1}{2}f_2\right) + f_{12}\left(P - M - \frac{1}{2}f_1\right)\right\}\omega^1$$

$$+ \left\{(f_{11} - f_{22})\left(P + \frac{1}{2}f_1\right) + f_{12}\left(Q - N + \frac{1}{2}f_2\right)\right\}\omega^2,$$

$$(6.34)\qquad \mathrm{d}\varphi(f) = -\{\Phi(f) + (\mathscr{L}f)^2 - 4\mathscr{M}f\}\,\omega^1 \wedge \omega^2,$$

$$\Phi(f) = \left(N + \frac{1}{2}f_2\right)\left(N - Q - \frac{1}{2}f_2\right) + \left(R + \frac{1}{2}f_1\right)\left(P - M - \frac{1}{2}f_1\right).$$

Further,

$$(6.35)\qquad \mathrm{d}\,\mathscr{L}f = \left(M + P + \frac{3}{2}f_1\right)\omega^1 + \left(N + Q + \frac{3}{2}f_2\right)\omega^2,$$

$$\mathrm{d}\,\mathscr{M}f = \left\{(f_{22} + f)\,M - 2f_{12}\left(N - \frac{1}{2}f_2\right)\right.$$

$$\left. + (f_{11} + f)\left(P - \frac{1}{2}f_1\right) + f_1\,\mathscr{L}f - 2f_2f_{12}\right\}\omega^1$$

$$+ \left\{(f_{22} + f)\left(N - \frac{1}{2}f_2\right) - 2f_{12}\left(P - \frac{1}{2}f_1\right)\right.$$

$$\left. + (f_{11} + f)\,Q + f_2\mathscr{L}f - 2f_1f_{22}\right\}\omega^2.$$

First of all, let us prove that the suppositions of our theorem imply $\Phi(f) \geqq 0$ in $\mathscr{D}$. Suppose (6.29). Then

$$(6.36)\qquad M + P + \frac{3}{2}f_1 = 0, \quad N + Q + \frac{3}{2}f_2 = 0,$$

and we have

$$(6.37)\qquad \Phi(f) = 4\left(N + \frac{1}{2}f_2\right)^2 + 4\left(P + \frac{1}{2}f_1\right)^2 \geqq 0.$$

Next, let

$$(6.38)\qquad \mathscr{M}f = 0 \quad \text{in } \mathscr{D}.$$

Then (6.35_2) implies

$$(6.39\qquad (f_{22}+f)\left(M-P+\frac{1}{2}f_1\right)+\mathscr{L}f\cdot\left(P+\frac{1}{2}f_1\right)-2f_{12}\left(N+\frac{1}{2}f_2\right)=0,$$

$$(f_{11}+f)\left(Q-N+\frac{1}{2}f_2\right)+\mathscr{L}f\cdot\left(N+\frac{1}{2}f_2\right)-2f_{12}\left(P+\frac{1}{2}f_1\right)=0.$$

Let $m \in \mathscr{D}$ be a fixed point; the frames may be always chosen in such a way that $f_{12}(m) = 0$. If $\mathscr{L}f(m) \neq 0$, we have

$$\{\Phi(f)\}(m) = 2\left\{\frac{f_{11}+f}{\mathscr{L}f}\left(N-Q-\frac{1}{2}f_2\right)^2+\frac{f_{22}+f}{\mathscr{L}f}\left(P-M-\frac{1}{2}f_1\right)^2\right\}_m.$$

Now, quite generally,

$$(f_{11}+f)\,\mathscr{L}f = \mathscr{M}f + f_{12}^2 + (f_{11}+f)^2,$$
$$(f_{22}+f)\,\mathscr{L}f = \mathscr{M}f + f_{12}^2 + (f_{22}+f)^2,$$

i.e., $\{\Phi(f)\}(m) \geqq 0$. In the case $\mathscr{L}f(m) = 0$, there are two possibilities: 1) $\mathscr{L}f = 0$ in a neighbourhood of m, 2) there is a sequence $\{m_i\}$, $m_i \to m$, such that $\mathscr{L}f(m_i) \neq 0$ for each m_i. The preceding results prove that $\{\Phi(f)\}(m) \geqq 0$ in these cases, too. Finally, consider the general supposition of our theorem. From (6.32) and (6.35), we get

$$(6.40)\qquad (f_{22}+f)\,M-2f_{12}\left(N-\frac{1}{2}f_2\right)+(f_{11}+f)\left(P-\frac{1}{2}f_1\right)$$
$$+f_1\mathscr{L}f-2f_2f_{12}-\frac{\mathrm{d}F}{\mathrm{d}t}\left(M+P+\frac{3}{2}f_1\right)=0,$$
$$(f_{22}+f)\left(N-\frac{1}{2}f_2\right)-2f_{12}\left(P-\frac{1}{2}f_1\right)+(f_{11}+f)\,Q$$
$$+f_2\mathscr{L}f-2f_1f_{12}-\frac{\mathrm{d}F}{\mathrm{d}t}\left(N+Q+\frac{3}{2}f_2\right)=0,$$

i.e.,

$$(6.41)\qquad \left(f_{22}+f-\frac{\mathrm{d}F}{\mathrm{d}t}\right)\left(M-P+\frac{1}{2}f_1\right)+\left(\mathscr{L}f-2\frac{\mathrm{d}F}{\mathrm{d}t}\right)\left(P+\frac{1}{2}f_1\right)$$
$$-2f_{12}\left(N+\frac{1}{2}f_2\right)=0,$$
$$\left(f_{11}+f-\frac{\mathrm{d}F}{\mathrm{d}t}\right)\left(Q-N+\frac{1}{2}f_2\right)+\left(\mathscr{L}f-2\frac{\mathrm{d}F}{\mathrm{d}t}\right)\left(N+\frac{1}{2}f_2\right)$$
$$-2f_{12}\left(P+\frac{1}{2}f_1\right)=0.$$

Suppose

$$\mathscr{L}f - 2\,\frac{\mathrm{d}F(\mathscr{L}f)}{\mathrm{d}t} = 0,$$

i.e., $\mathscr{M}f = F(\mathscr{L}f) = \frac{1}{4}(\mathscr{L}f)^2 + c$, $c = \text{const}$. The condition (6.31) implies $\frac{1}{4}t^2 + c > \frac{1}{2}t\left(t - \frac{1}{2}t\right)$, i.e., $c > 0$. On the other hand, (6.28) implies $-4c = (\mathscr{L}f)^2 - 4\mathscr{M}f \geqq 0$, which is a contradiction. Thus

$$\mathscr{L}f - 2\,\frac{\mathrm{d}F(\mathscr{L}f)}{\mathrm{d}t} \neq 0 \quad \text{in } \mathscr{D}.$$

Let $m \in \mathscr{D}$ be again a point, and suppose $f_{12}(m) = 0$. Then

$$\{\Phi(f)\}(m) = 2\left(\mathscr{L}f - 2\,\frac{\mathrm{d}F}{\mathrm{d}t}\right)^{-1} \times \left\{\left(f_{11} + f - \frac{\mathrm{d}F}{\mathrm{d}t}\right)\left(N - Q - \frac{1}{2}f_2\right)^2 + \left(f_{22} + f - \frac{\mathrm{d}F}{\mathrm{d}t}\right)\left(P - M - \frac{1}{2}f_1\right)^2\right\}_m .$$

It is easy to verify

$$\left(f_{11} + f - \frac{\mathrm{d}F}{\mathrm{d}t}\right)\left(\mathscr{L}f - 2\,\frac{\mathrm{d}F}{\mathrm{d}t}\right) = \left(\frac{\mathrm{d}F}{\mathrm{d}t}\right)^2 - \frac{\mathrm{d}F}{\mathrm{d}t}\cdot\mathscr{L}f + \mathscr{M}f + \left(f_{11} + f - \frac{\mathrm{d}F}{\mathrm{d}t}\right)^2,$$

$$\left(f_{22} + f - \frac{\mathrm{d}F}{\mathrm{d}t}\right)\left(\mathscr{L}f - 2\,\frac{\mathrm{d}F}{\mathrm{d}t}\right) = \left(\frac{\mathrm{d}F}{\mathrm{d}t}\right)^2 - \frac{\mathrm{d}F}{\mathrm{d}t}\cdot\mathscr{L}f + \mathscr{M}f + \left(f_{22} + f - \frac{\mathrm{d}F}{\mathrm{d}t}\right)^2;$$

because of (6.32) and (6.31),

$$\left(f_{11} + f - \frac{\mathrm{d}F}{\mathrm{d}t}\right)\left(\mathscr{L}f - 2\,\frac{\mathrm{d}F}{\mathrm{d}t}\right) > 0, \quad \left(f_{22} + f - \frac{\mathrm{d}F}{\mathrm{d}t}\right)\left(\mathscr{L}f - 2\,\frac{\mathrm{d}F}{\mathrm{d}t}\right) > 0,$$

and $\{\Phi(f)\}(m) \geqq 0$ follows.

By means of (6.28), we get

(6.42) $\quad f_{11} - f_{22} = f_{12} = 0 \quad \text{on } \partial\mathscr{D}$

in all cases. Thus $\varphi(f) = 0$ on $\partial\mathscr{D}$, and we get

(6.43) $\quad f_{11} - f_{22} = f_{12} = 0 \quad \text{in } \mathscr{D}$

from the Stokes' formula for $\varphi(f)$. From this and (6.29) or (6.32) resp., we obtain

$$f_{11} = -f, \quad f_{22} = -f, \quad f_{12} = 0 \quad \text{in } \mathscr{D}. \tag{6.44}$$

Now, consider the vector field

$$a = -f_1 v_1 - f_2 v_2 + f v_3. \tag{6.45}$$

Then $\mathrm{d}a = 0$, i.e., $a = \text{const}$, and $f = \langle v_3, a \rangle$. QED.

Let M be an ovaloid in R^3, i.e., let $\partial M = \emptyset$ and $K > 0$. Be given a function $f\colon M \to \mathbf{R}$. Introduce the (obviously invariant) function

$$F = \frac{1}{2}(f_{11}^2 + 2f_{12}^2 + f_{22}^2). \tag{6.46}$$

Then it is easy to see that

$$\begin{aligned} \mathrm{d}F = {} & \left(f_{11}M + 2f_{12}N + f_{22}P + Kf_2 f_{12} - \frac{1}{2}Kf_1 f_{22}\right)\omega^1 \\ & + \left(f_{11}N + 2f_{12}P + f_{22}Q + Kf_1 f_{12} - \frac{1}{2}Kf_2 f_{11}\right)\omega^2, \end{aligned} \tag{6.47}$$

$$\begin{aligned} \Delta F = {} & M^2 + 3N^2 + 3P^2 + Q^2 + K(f_1 P + f_2 N) \\ & + f_{11}R + 2f_{12}S + (f_{11} + f_{22})\,T + 2f_{12}U + f_{22}V \\ & + K\left(\frac{3}{2}f_{11}^2 - f_{11}f_{22} + \frac{3}{2}f_{22}^2 + 7f_{12}^2\right) + \frac{1}{2}(K_1 f_1 - K_2 f_2)(f_{11} - f_{22}) \\ & + \frac{3}{2}(K_1 f_2 + K_2 f_1)\,f_{12} + \frac{3}{4}K^2(f_1^2 + f_2^2). \end{aligned} \tag{6.48}$$

Further,

$$\Delta f^2 = 2\{f_1^2 + f_2^2 + f(f_{11} + f_{22})\}. \tag{6.49}$$

Let us suppose that f is a solution of the partial differential equation

$$\Delta f + 2cf = 0, \qquad c = \text{const}, \tag{6.50}$$

on M. From (6.50), we get

$$M + P + \left(2c - \frac{1}{2}K\right)f_1 = 0, \quad N + Q + \left(2c - \frac{1}{2}K\right)f_2 = 0; \tag{6.51}$$

$$\begin{aligned} & R + T + \left(2c - \frac{1}{2}K\right)f_{11} + \frac{3}{2}Kf_{22} - \frac{1}{2}K_1 f_1 + \frac{1}{2}K_2 f_2 = 0, \\ & S + U + \left(2c - \frac{1}{2}K\right)f_{12} - \frac{1}{4}K_2 f_1 - \frac{1}{4}K_1 f_2 = 0, \\ & T + V + \frac{3}{2}Kf_{11} + \left(2c - \frac{1}{2}K\right)f_{22} + \frac{1}{2}K_1 f_1 - \frac{1}{2}K_2 f_2 = 0 \end{aligned} \tag{6.52}$$

and the existence of functions φ and ψ such that

$$f_{11} = \varphi - cf, \quad f_{22} = -\varphi - cf; \tag{6.53}$$

$$M = -P + \left(\frac{1}{2}K - 2c\right) f_1, \quad Q = -N + \left(\frac{1}{2}K - 2c\right) f_2;$$

$$R = -T + \frac{1}{2}K_1 f_1 - \frac{1}{2}K_2 f_2 + 2(K - c)\varphi + c(K + 2c) f,$$

$$S = \psi + \frac{1}{8}(K_1 f_2 + K_2 f_1) + \left(\frac{1}{4}K - c\right) f_{12},$$

$$U = -\psi + \frac{1}{8}(K_1 f_2 + K_2 f_1) + \left(\frac{1}{4}K - c\right) f_{12},$$

$$V = -T - \frac{1}{2}K_1 f_1 + \frac{1}{2}K_2 f_2 - 2(K - c)\varphi + c(K + 2c) f.$$

Using these and (6.48), (6.49), we get

$$\begin{aligned} \Delta(F - c^2 f^2) = {} & (2P + cf_1)^2 + (2N + cf_2)^2 \\ & + 4(2K - c)(f_{12}^2 + \varphi^2) + 2(K_1 f_2 + K_2 f_1) f_{12} \\ & + 2(K_1 f_1 - K_2 f_2)\varphi + (K - c)^2 (f_1^2 + f_2^2). \end{aligned} \tag{6.54}$$

The integral formula

$$\int_M \Delta(F - c^2 f^2)\, \omega^1 \wedge \omega^2 = 0 \tag{6.55}$$

enables us to prove

Theorem 6.3. *Let M be a sphere with the Gauss curvature K and let $f\colon M \to \mathbf{R}$ be a solution of* (6.50). *If $K = c$, f is linear. If $c \neq K > \frac{1}{2}c$, $f =$* const.

Proof. For $K = c$, (6.54) reduces to

$$\Delta(F - K^2 f^2) = (2P + Kf_1)^2 + (2N + Kf_2)^2 + 4K(f_{12}^2 + \varphi^2), \tag{6.56}$$

and we get

$$f_{12} = 0, \qquad \varphi = \frac{1}{2}(f_{11} - f_{22}) = 0$$

from (6.55), i.e., we have (6.43). For $c \neq K > \frac{1}{2}c$, we get $f_1 = f_2 = 0$, i.e., $f =$ const. QED.

Theorem 6.4. *Let M be an ovaloid, $c \in \mathbf{R}$, and let the Gauss curvature K of M satisfy*

$$K > \frac{1}{2}c. \tag{6.57}$$

Let the function $f\colon M \to \mathbf{R}$ satisfy (6.50). (i) *Let f have a finite number of local maximal and minimal values. Then M is a sphere with $K = c$.* (ii) *Let us suppose that K does* *not attain its local maximum or minimum at the points $m \in M$ such that $K(m) = c$. Then $f = \text{const}$.*

Proof. Because of (6.57), (6.54) may be rewritten as

$$(6.58) \quad \Delta(F - c^2 f^2) = (2P + cf_1)^2 + (2N + cf_2)^2 + \frac{1}{4(2K - c)} \{4(2K - c) f_{12} + K_1 f_2 + K_2 f_1\}^2 + \frac{1}{4(2K - c)} \{4(2K - c) \varphi + K_1 f_1 - K_2 f_2\}^2 + \frac{1}{4(2K - c)} \{4(2K - c) (K - c)^2 + K_1^2 + K_2^2\} (f_1^2 + f_2^2).$$

(i) In this case, $f_1^2 + f_2^2 \neq 0$ with the exception of a finite set S of points of M. Thus $K = c$ on $M - S$, from the continuity, $K = c$ on M. (ii) The supposition reads $4(2K - c)(K - c)^2 + K_1^2 + K_2^2 > 0$ on M, and we get $f_1 = f_2 = 0$ on M. QED.

VI. Global differential geometry of isometries

1. General theory of infinitesimal isometries

Let $M: \mathscr{D} \to E^3$ be a surface. To each $t \in \mathbf{R}$ in some (possibly small) neighbourhood I of $0 \in \mathbf{R}$, associate a surface $M_t: \mathscr{D} \to E^3$ such that (i) $M_0 = M$, (ii) M_t depends analytically on t in the sense that the mapping $I_d: I \to E^3$, $I_d(t) = M_t(d)$, is of class C^ω for each point $d \in \mathscr{D}$. To each M_t, associate the frames $\{m(t), v_1(t), v_2(t), v_3(t)\}$ in such a way that $v_i(t)$ depend analytically on t, too; we are then in the position to write

$$
\begin{aligned}
(1.1)\qquad \mathrm{d}m(t) &= (\omega^1 + t\varphi^1 + \cdots)\, v_1(t) + (\omega^2 + t\varphi^2 + \cdots)\, v_2(t),\\
\mathrm{d}v_1(t) &= (\omega_1{}^2 + t\varphi_1{}^2 + \cdots)\, v_2(t) + (\omega_1{}^3 + t\varphi_1{}^3 + \cdots)\, v_3(t),\\
\mathrm{d}v_2(t) &= -(\omega_1{}^2 + t\varphi_1{}^2 + \cdots)\, v_1(t) + (\omega_2{}^3 + t\varphi_2{}^3 + \cdots)\, v_3(t),\\
\mathrm{d}v_3(t) &= -(\omega_1{}^3 + t\varphi_1{}^3 + \cdots)\, v_1(t) - (\omega_2{}^3 + t\varphi_2{}^3 + \cdots)\, v_2(t)
\end{aligned}
$$

with the integrability conditions

$$
\begin{aligned}
(1.2)\qquad & (\omega^1 + t\varphi^1 + \cdots) \wedge (\omega_1{}^3 + t\varphi_1{}^3 + \cdots) + (\omega^2 + t\varphi^2 + \cdots) \wedge (\omega_2{}^3 + t\varphi_2{}^3 + \cdots)\\
&= 0,\\
& \mathrm{d}(\omega^1 + t\varphi^1 + \cdots) = -(\omega^2 + t\varphi^2 + \cdots) \wedge (\omega_1{}^2 + t\varphi_1{}^2 + \cdots),\\
& \mathrm{d}(\omega^2 + t\varphi^2 + \cdots) = (\omega^1 + t\varphi^1 + \cdots) \wedge (\omega_1{}^2 + t\varphi_1{}^2 + \cdots),\\
& \mathrm{d}(\omega_1{}^2 + t\varphi_1{}^2 + \cdots) = -(\omega_1{}^3 + t\varphi_1{}^3 + \cdots) \wedge (\omega_2{}^3 + t\varphi_2{}^3 + \cdots),\\
& \mathrm{d}(\omega_1{}^3 + t\varphi_1{}^3 + \cdots) = (\omega_1{}^2 + t\varphi_1{}^2 + \cdots) \wedge (\omega_2{}^3 + t\varphi_2{}^3 + \cdots),\\
& \mathrm{d}(\omega_2{}^3 + t\varphi_2{}^3 + \cdots) = -(\omega_1{}^2 + t\varphi_1{}^2 + \cdots) \wedge (\omega_1{}^3 + t\varphi_1{}^3 + \cdots).
\end{aligned}
$$

Comparing the terms at t, we get

$$(1.3)\qquad \omega^1 \wedge \varphi_1{}^3 + \varphi^1 \wedge \omega_1{}^3 + \omega^2 \wedge \varphi_2{}^3 + \varphi^2 \wedge \omega_2{}^3 = 0,$$

$$(1.4)\qquad \mathrm{d}\varphi^1 = -\omega^2 \wedge \varphi_1{}^2 - \varphi^2 \wedge \omega_1{}^3, \qquad \mathrm{d}\varphi^2 = \omega^1 \wedge \varphi_1{}^3 + \varphi^1 \wedge \omega_1{}^2,$$

$$(1.5)\qquad \mathrm{d}\varphi_1{}^2 = -\omega_1{}^3 \wedge \varphi_2{}^3 - \varphi_1{}^3 \wedge \omega_2{}^3,$$

$$(1.6)\qquad \mathrm{d}\varphi_1{}^3 = \omega_1{}^2 \wedge \varphi_2{}^3 + \varphi_1{}^2 \wedge \omega_2{}^3, \quad \mathrm{d}\varphi_2{}^3 = -\omega_1{}^2 \wedge \varphi_1{}^3 - \varphi_1{}^2 \wedge \omega_1{}^3.$$

Definition 1.1. *Let $M: \mathscr{D} \to E^3$ be a surface and $\{m, v_1, v_2, v_3\}$ a field of associated frames such that we have* (IV.2.1). *Each set Φ of forms $(\varphi^1, \varphi^2, \varphi_1{}^2, \varphi_1{}^3, \varphi_2{}^3)$ satisfying* (1.3)—(1.6) *is called the* infinitesimal deformation *of the surface M.*

This definition is not very attractive; it is too much analytic and non-invariant. Because of this, I am going to present a formally more special definition without entering into the details.

Be given a surface $M: \mathscr{D} \to E^3$. At each point $m \in M$, be given a vector $v(m)$ starting at m, and let us consider the layer of surfaces M_t, $m(t) = m + t \cdot v(m)$; we have to restrict ourselves to t's in a neighbourhood of $0 \in \mathbf{R}$ in order to ensure that all M_t's are surfaces. Let us evaluate equations (1.1) in this special case. The vector field v be given by

$$v = xv_1 + yv_2 + zv_3. \tag{1.7}$$

The covariant derivatives $x_i, y_i, z_i, x_{ij} = x_{ji}, y_{ij} = y_{ji}, z_{ij} = z_{ji}, x_{ijk}, y_{ijk}, z_{ijk}$ be introduced by the usual prolongation procedure:

$$\begin{aligned} &dx - y\omega_1{}^2 = x_1\omega^1 + x_2\omega^2, \\ &dy + x\omega_1{}^2 = y_1\omega^1 + y_2\omega^2, \\ &dz = z_1\omega^1 + z_2\omega^2; \end{aligned} \tag{1.8}$$

$$\begin{aligned} &\{dx_1 - (x_2 + y_1)\,\omega_1{}^2\} \wedge \omega^1 + \{dx_2 + (x_1 - y_2)\,\omega_1{}^2\} \wedge \omega^2 = Ky\omega^1 \wedge \omega^2, \\ &\{dy_1 + (x_1 - y_2)\,\omega_1{}^2\} \wedge \omega^1 + \{dy_2 + (x_2 + y_1)\,\omega_1{}^2\} \wedge \omega^2 = -Kx\omega^1 \wedge \omega^2, \\ &(dz_1 - z_2\omega_1{}^2) \wedge \omega^1 + (dz_2 + z_1\omega_1{}^2) \wedge \omega^2 = 0; \end{aligned} \tag{1.9}$$

$$\begin{aligned} &dx_1 - (x_2 + y_1)\,\omega_1{}^2 = x_{11}\omega^1 + \left(x_{12} - \frac{1}{2}\,Ky\right)\omega^2, \\ &dx_2 + (x_1 - y_2)\,\omega_1{}^2 = \left(x_{12} + \frac{1}{2}\,Ky\right)\omega^1 + x_{22}\omega^2, \\ &dy_1 + (x_1 - y_2)\,\omega_1{}^2 = y_{11}\omega^1 + \left(y_{12} + \frac{1}{2}\,Kx\right)\omega^2, \\ &dy_2 + (x_2 + y_1)\,\omega_1{}^2 = \left(y_{12} - \frac{1}{2}\,Kx\right)\omega^1 + y_{22}\omega^2, \\ &dz_1 - z_2\omega_1{}^2 = z_{11}\omega^1 + z_{12}\omega^2, \\ &dz_2 + z_1\omega_1{}^2 = z_{12}\omega^1 + z_{22}\omega^2; \end{aligned} \tag{1.10}$$

$$\begin{aligned} &\{dx_{11} - (2x_{12} + y_{11})\,\omega_1{}^2\} \wedge \omega^1 + \{dx_{12} + (x_{11} - x_{22} - y_{12})\,\omega_1{}^2\} \wedge \omega^2 \\ &= \left(Kx_2 + \frac{3}{2}\,Ky_1 + \frac{1}{2}\,K_1y\right)\omega^1 \wedge \omega^2, \\ &\{dx_{12} + (x_{11} - x_{22} - y_{12})\,\omega_1{}^2\} \wedge \omega^1 + \{dx_{22} + (2x_{12} - y_{22})\,\omega_1{}^2\} \wedge \omega^2 \\ &= \left(-Kx_1 + \frac{3}{2}\,Ky_2 + \frac{1}{2}\,K_2y\right)\omega^1 \wedge \omega^2, \\ &\{dy_{11} - (2y_{12} - x_{11})\,\omega_1{}^2\} \wedge \omega^1 + \{dy_{12} + (y_{11} - y_{22} + x_{12})\,\omega_1{}^2\} \wedge \omega^2 \\ &= \left(-\frac{3}{2}\,Kx_1 + Ky_2 - \frac{1}{2}\,K_1x\right)\omega^1 \wedge \omega^2, \\ &\{dy_{12} + (y_{11} - y_{22} + x_{12})\,\omega_1{}^2\} \wedge \omega^1 + \{dy_{22} + (2y_{12} + x_{22})\,\omega_1{}^2\} \wedge \omega^2 \\ &= \left(-\frac{3}{2}\,Kx_2 - Ky_1 - \frac{1}{2}\,K_2x\right)\omega^1 \wedge \omega^2, \\ &(dz_{11} - 2z_{12}\omega_1{}^2) \wedge \omega^1 + \{dz_{12} + (z_{11} - z_{22})\,\omega_1{}^2\} \wedge \omega^2 = Kz_2\omega^1 \wedge \omega^2, \\ &\{dz_{12} + (z_{11} - z_{22})\,\omega_1{}^2\} \wedge \omega^1 + (dz_{22} + 2z_{12}\omega_1{}^2) \wedge \omega^2 = -Kz_1\omega^1 \wedge \omega^2; \end{aligned} \tag{1.11}$$

$$
\begin{aligned}
(1.12)\quad & dx_{11} - (2x_{12} + y_{11})\,\omega_1{}^2 = x_{111}\omega^1 + x_{112}\omega^2,\\
& dx_{12} + (x_{11} - x_{22} - y_{12})\,\omega_1{}^2 = x_{121}\omega^1 + x_{122}\omega^2,\\
& dx_{22} + (2x_{12} - y_{22})\,\omega_1{}^2 = x_{221}\omega^1 + x_{222}\omega^2,\\
& dy_{11} - (2y_{12} - x_{11})\,\omega_1{}^2 = y_{111}\omega^1 + y_{112}\omega^2,\\
& dy_{12} + (y_{11} - y_{22} + x_{12})\,\omega_1{}^2 = y_{121}\omega^1 + y_{122}\omega^2,\\
& dy_{22} + (2y_{12} + x_{22})\,\omega_1{}^2 = y_{221}\omega^1 + y_{222}\omega^2,\\
& dz_{11} - 2z_{12}\omega_1{}^2 = z_{111}\omega^1 + z_{112}\omega^2,\\
& dz_{12} + (z_{11} - z_{22})\,\omega_1{}^2 = z_{121}\omega^1 + z_{122}\omega^2,\\
& dz_{22} + 2z_{12}\omega_1{}^2 = z_{221}\omega^1 + z_{222}\omega^2,
\end{aligned}
$$

where $dK = K_1\omega^1 + K_2\omega^2$ and

$$
\begin{aligned}
(1.13)\quad & x_{121} - x_{112} = Kx_2 + \frac{3}{2}\,Ky_1 + \frac{1}{2}\,K_1y,\\
& x_{221} - x_{122} = -Kx_1 + \frac{3}{2}\,Ky_2 + \frac{1}{2}\,K_2y,\\
& y_{121} - y_{112} = -\frac{3}{2}\,Kx_1 + Ky_2 - \frac{1}{2}\,K_1x,\\
& y_{221} - y_{122} = -\frac{3}{2}\,Kx_2 - Ky_1 - \frac{1}{2}\,K_2x,\\
& z_{121} - z_{112} = Kz_2, \qquad z_{221} - z_{122} = -Kz_1.
\end{aligned}
$$

To each surface M_t, associate the orthonormal frames

$$
\begin{aligned}
(1.14)\quad & m(t) = m + t(xv_1 + yv_2 + zv_3), \quad v_1(t) = v_1 + t\cdot\varrho_1v_3 + \cdots,\\
& v_2(t) = v_2 + t\cdot\varrho_2v_3 + \cdots, \quad v_3(t) = v_3 - t\cdot(\varrho_1v_1 + \varrho_2v_2) + \cdots.
\end{aligned}
$$

Inserting this into (1.1), we easily obtain

$$
(1.15)\quad \varrho_1 = z_1 + ax + by, \qquad \varrho_2 = z_2 + bx + cy,
$$

$$
\begin{aligned}
(1.16)\quad & \varphi^1 = (x_1 - az)\,\omega^1 + (x_2 - bz)\,\omega^2, \qquad \varphi^2 = (y_1 - bz)\,\omega^1 + (y_2 - cz)\,\omega^2,\\
& \varphi_1{}^2 = (az_2 - bz_1 + Ky)\,\omega^1 + (bz_2 - cz_1 - Kx)\,\omega^2,\\
& \varphi_1{}^3 = (z_{11} + ax_1 + by_1 + \alpha x + \beta y)\,\omega^1 + (z_{12} + ax_2 + by_2 + \beta x + \gamma y)\,\omega^2,\\
& \varphi_2{}^3 = (z_{12} + bx_1 + cy_1 + \beta x + \gamma y)\,\omega^1 + (z_{22} + bx_2 + cy_2 + \gamma x + \delta y)\,\omega^2,
\end{aligned}
$$

$a, \ldots, c, \alpha, \ldots, \delta$ being introduced by (IV.2.22) and (IV.2.24). The following lemma is obvious (or may be proved by a direct calculation).

Lemma 1.1. *The set of forms* (1.16) *is an infinitesimal deformation of the surface M.*

Using the above construction, we have geometrically obtained, from a given vector field v, an infinitesimal deformation of our surface. And the other way round, it would be possible to show that each infinitesimal deformation might be (at least locally) constructed this way.

The *variations* of the fundamental forms of the surface with respect to a given infinitesimal deformation are given by

$$(1.17)\quad \begin{aligned} I + t\cdot\delta I + \cdots &= (\omega^1 + t\varphi^1 + \cdots)^2 + (\omega^2 + t\varphi^2 + \cdots)^2,\\ II + t\cdot\delta II + \cdots &= (\omega^1 + t\varphi^1 + \cdots)(\omega_1{}^3 + t\varphi_1{}^3 + \cdots)\\ &\quad + (\omega^2 + t\varphi^2 + \cdots)(\omega_2{}^3 + t\varphi_2{}^3 + \cdots),\\ III + t\cdot\delta III + \cdots &= (\omega_1{}^3 + t\varphi_1{}^3 + \cdots)^2 + (\omega_2{}^3 + t\varphi_2{}^3 + \cdots)^2, \end{aligned}$$

i.e.,

$$(1.18)\quad \begin{aligned} \delta I &= 2(\omega^1\varphi^1 + \omega^2\varphi^2),\\ \delta II &= \omega_1\varphi_1{}^3 + \varphi^1\omega_1{}^3 + \omega^2\varphi_2{}^3 + \varphi^2\omega_2{}^3,\\ \delta III &= 2(\omega_1{}^3\varphi_1{}^3 + \omega_2{}^3\varphi_2{}^3). \end{aligned}$$

Notice that the curvatures of M may be defined by

$$K\omega^1 \wedge \omega^2 = \omega_1{}^3 \wedge \omega_2{}^3 \quad\text{and}\quad 2H\omega^1 \wedge \omega^2 = \omega_1{}^3 \wedge \omega^2 + \omega^1 \wedge \omega_2{}^3$$

resp. Thus it is possible to define the *variations* δH and δK by means of

$$(1.19)\quad \begin{aligned} &(K + t\cdot\delta K + \cdots)(\omega^1 + t\varphi^1 + \cdots) \wedge (\omega^2 + t\varphi^2 + \cdots)\\ &= (\omega_1{}^3 + t\varphi_1{}^3 + \cdots) \wedge (\omega_2{}^3 + t\varphi_2{}^3 + \cdots),\\ &2(H + t\cdot\delta H + \cdots)(\omega^1 + t\varphi^1 + \cdots) \wedge (\omega^2 + t\varphi^2 + \cdots)\\ &= (\omega_1{}^3 + t\varphi_1{}^3 + \cdots) \wedge (\omega^2 + t\varphi^2 + \cdots)\\ &\qquad + (\omega^1 + t\varphi^1 + \cdots) \wedge (\omega_2{}^3 + t\varphi_2{}^3 + \cdots), \end{aligned}$$

and we obtain

$$(1.20)\quad \begin{aligned} 2\delta H\cdot\omega^1 \wedge \omega^2 &= \omega_1{}^3 \wedge \varphi^2 + \varphi_1{}^3 \wedge \omega^2 + \omega^1 \wedge \varphi_2{}^3 + \varphi^1 \wedge \omega_2{}^3\\ &\qquad - 2H(\omega^1 \wedge \varphi^2 + \varphi^1 \wedge \omega^2),\\ \delta K\cdot\omega^1 \wedge \omega^2 &= \omega_1{}^3 \wedge \varphi_2{}^3 + \varphi_1{}^3 \wedge \omega_2{}^3 - K(\omega^1 \wedge \varphi^2 + \varphi^1 \wedge \omega^2). \end{aligned}$$

Finally, let us study the possible changes of the frames in (1.1). Let

$$(1.21)\quad \begin{aligned} v_1(t) &= \varepsilon_1 \cos(\varrho + t\sigma + \cdots)\cdot {}^*v_1(t) - \varepsilon_1 \sin(\varrho + t\sigma + \cdots)\cdot {}^*v_2(t),\\ v_2(t) &= \sin(\varrho + t\sigma + \cdots)\cdot {}^*v_1(t) + \cos(\varrho + t\sigma + \cdots)\cdot {}^*v_2(t),\\ v_3(t) &= \varepsilon_2 {}^*v_3(t); \qquad \varepsilon_1{}^2 = \varepsilon_2{}^2 = 1; \end{aligned}$$

of course,

$$(1.22)\quad \begin{aligned} \sin(\varrho + t\sigma + \cdots) &= \sin\varrho + t\cdot\sigma\cos\varrho + \cdots,\\ \cos(\varrho + t\sigma + \cdots) &= \cos\varrho - t\cdot\sigma\sin\varrho + \cdots. \end{aligned}$$

Inserting this into (1.1) and analogous equations (*1.1), we get

$$(1.23)\quad \begin{aligned} &{}^*\omega^1 = \varepsilon_1\cos\varrho\cdot\omega^1 + \sin\varrho\cdot\omega^2, \quad {}^*\omega^2 = -\varepsilon_1\sin\varrho\cdot\omega^1 + \cos\varrho\cdot\omega^2,\\ &{}^*\omega_1{}^2 = \varepsilon_1\omega_1{}^2 + d\varrho,\\ &{}^*\omega_1{}^3 = \varepsilon_1\varepsilon_2\cos\varrho\cdot\omega_1{}^3 + \varepsilon_2\sin\varrho\cdot\omega_2{}^3,\\ &{}^*\omega_2{}^3 = -\varepsilon_1\varepsilon_2\sin\varrho\cdot\omega_1{}^3 + \varepsilon_2\cos\varrho\cdot\omega_2{}^3 \end{aligned}$$

and

$$(1.24)\quad \begin{aligned} {}^*\varphi^1 &= -\varepsilon_1\sigma \sin\varrho\cdot\omega^1 + \sigma\cos\varrho\cdot\omega^2 + \varepsilon_1\cos\varrho\cdot\varphi^1 + \sin\varrho\cdot\varphi^2,\\ {}^*\varphi^2 &= -\varepsilon_1\sigma\cos\varrho\cdot\omega^1 - \sigma\sin\varrho\cdot\omega^2 - \varepsilon_1\sin\varrho\cdot\varphi^1 + \cos\varrho\cdot\varphi^2,\\ {}^*\varphi_1{}^2 &= \varepsilon_1\varphi_1{}^2 + \mathrm{d}\sigma,\\ {}^*\varphi_1{}^3 &= \varepsilon_1\varepsilon_2\cos\varrho\cdot\varphi_1{}^3 + \varepsilon_2\sin\varrho\cdot\varphi_2{}^3 - \varepsilon_1\varepsilon_2\sigma\sin\varrho\cdot\omega_1{}^3 + \varepsilon_2\sigma\cos\varrho\cdot\omega_2{}^3,\\ {}^*\varphi_2{}^3 &= -\varepsilon_1\varepsilon_2\sin\varrho\cdot\varphi_1{}^3 + \varepsilon_2\cos\varrho\cdot\varphi_2{}^3 - \varepsilon_1\varepsilon_2\sigma\cos\varrho\cdot\omega_1{}^3 - \varepsilon_2\sigma\sin\varrho\cdot\omega_2{}^3. \end{aligned}$$

Definition 1.2. *The infinitesimal deformation Φ satisfying $\delta I = 0$ is called the* infinitesimal I-isometry; *similarly, we define the* infinitesimal II- *and* III-isometries *resp. The surface M is said to be* infinitesimally rigid with respect to Φ *if* $\delta I = \delta II = 0$.

2. I-isometries

First of all, let us study the infinitesimal I-isometries of a given surface M.

Theorem 2.1. *Let $M: \mathscr{D} \to E^3$, $\mathscr{D} \subset \mathbf{R}^2$ a bounded domain, be a surface and Φ its infinitesimal I-isometry. Suppose:* (i) $K > 0$ *on* M; (ii) *there is a function $\lambda: \partial M \to \mathbf{R}$ such that $\delta II = \lambda \cdot I$ on ∂M. Then M is infinitesimally rigid with respect to Φ.*

Proof. Let us write

$$(2.1)\quad \varphi^1 = a_1\omega^1 + a_2\omega^2, \qquad \varphi^2 = a_3\omega^1 + a_4\omega^2;$$

then, see (1.18_1),

$$(2.2)\quad \frac{1}{2}\,\delta I = a_1(\omega^1)^2 + (a_2 + a_3)\,\omega^1\omega^2 + a_4(\omega^2)^2.$$

Because of $\delta I = 0$, there is $a_1 = a_2 + a_3 = a_4 = 0$, i.e., the forms φ_1, φ^2 take the form

$$(2.3)\quad \varphi^1 = a_0\omega^2, \qquad \varphi^2 = -a_0\omega^1.$$

Inserting this into (1.24), we get

$$(2.4)\quad \begin{aligned} {}^*\varphi^1 &= (\varepsilon_1\sigma + A)\,(-\sin\varrho\cdot\omega^1 + \varepsilon_1\cos\varrho\cdot\omega^2),\\ {}^*\varphi^2 &= (\varepsilon_1\sigma + A)\,(-\cos\varrho\cdot\omega^1 - \varepsilon_1\sin\varrho\cdot\omega^2). \end{aligned}$$

For $\sigma = -\varepsilon_1 A$, ${}^*\varphi^1 = {}^*\varphi^2 = 0$; thus we are in the position to choose the frames in such a way that

$$(2.5)\quad \varphi^1 = \varphi^2 = 0.$$

From (1.4),

$$(2.6)\quad \varphi_1{}^2 = 0.$$

From (1.3), we get the existence of functions R_1, R_2, R_3 such that

$$(2.7)\quad \varphi_1{}^3 = R_1\omega^1 + R_2\omega^2, \qquad \varphi_2{}^3 = R_2\omega^1 + R_3\omega^2;$$

(2.6), (2.7) and (1.5) yield

$$cR_1 - 2bR_2 + aR_3 = 0. \tag{2.8}$$

From $H^2 \geqq K > 0$, $H \neq 0$ on M, and (2.8) may be rewritten as

$$R_1 = H^{-1}\left\{\frac{1}{2}\,a(R_1 - R_3) + bR_2\right\}, \qquad R_3 = -H^{-1}\left\{\frac{1}{2}\,c(R_1 - R_3) - bR_2\right\}. \tag{2.9}$$

From (2.7) and (1.6), we get the existence of functions $S_1, \ldots, S_4$ such that

$$\begin{aligned} &dR_1 - 2R_2\omega_1{}^2 = S_1\omega^1 + S_2\omega^2,\\ &dR_2 + (R_1 - R_3)\,\omega_1{}^2 = S_2\omega^1 + S_3\omega^2,\\ &dR_3 + 2R_2\omega_1{}^2 = S_3\omega^1 + S_4\omega^2 \end{aligned} \tag{2.10}$$

which implies

$$\begin{aligned} &d(R_1 - R_3) - 4R_2\omega_1{}^2 = (S_1 - S_3)\,\omega^1 + (S_2 - S_4)\,\omega^2,\\ &dR_2 + (R_1 - R_3)\,\omega_1{}^2 = S_2\omega^1 + S_3\omega^2. \end{aligned} \tag{2.11}$$

The differential consequences of (2.8) are

$$\begin{aligned} &cS_1 - 2bS_2 + aS_3 = -\gamma R_1 + 2\beta R_2 - \alpha R_3,\\ &cS_2 - 2bS_3 + aS_4 = -\delta R_1 + 2\gamma R_2 - \beta R_3. \end{aligned} \tag{2.12}$$

Now, let us introduce, on $\mathscr{D}$, coordinates (u, v) such that (V.1.4) and (V.1.5) hold true. Then (2.11) read

$$\begin{aligned} &\frac{\partial(R_1 - R_3)}{\partial u} + \frac{4}{s}\cdot\frac{\partial r}{\partial v}\,R_2 = (S_1 - S_3)\,r,\\ &\frac{\partial(R_1 - R_3)}{\partial v} - \frac{4}{r}\,\frac{\partial s}{\partial u}\,R_2 = (S_2 - S_4)\,s,\\ &\frac{\partial R_2}{\partial u} - \frac{1}{s}\,\frac{\partial r}{\partial v}\,(R_1 - R_3) = S_2 r,\\ &\frac{\partial R_2}{\partial v} + \frac{1}{r}\,\frac{\partial s}{\partial u}\,(R_1 - R_3) = S_3 s. \end{aligned} \tag{2.13}$$

Thus

$$\begin{aligned} &rsS_1 = r\,\frac{\partial R_2}{\partial v} + s\,\frac{\partial(R_1 - R_3)}{\partial u} + 4\,\frac{\partial r}{\partial v}\,R_2 + \frac{\partial s}{\partial u}\,(R_1 - R_3),\\ &rsS_2 = s\,\frac{\partial R_2}{\partial u} - \frac{\partial r}{\partial v}\,(R_1 - R_3),\\ &rsS_3 = r\,\frac{\partial R_2}{\partial v} + \frac{\partial s}{\partial u}\,(R_1 - R_3),\\ &rsS_4 = s\,\frac{\partial R_2}{\partial u} - r\,\frac{\partial(R_1 - R_3)}{\partial v} + 4\,\frac{\partial s}{\partial u}\,R_2 - \frac{\partial r}{\partial v}\,(R_1 - R_3). \end{aligned} \tag{2.14}$$

Because of (2.9), the functions R_1 and R_3 are linear combinations of the functions $R_1 - R_3$ and R_2 resp. Multiplying (2.12) by rs and inserting there (2.14), we get the system

$$(2.15)\qquad -2bs\frac{\partial R_2}{\partial u} + 2Hr\frac{\partial R_2}{\partial v} + cs\frac{\partial(R_1 - R_3)}{\partial u} = c_{11}R_2 + c_{12}(R_1 - R_3),$$

$$2Hs\frac{\partial R_2}{\partial u} - 2br\frac{\partial R_2}{\partial v} - ar\frac{\partial(R_1 - R_3)}{\partial v} = c_{21}R_2 + c_{22}(R_1 - R_3),$$

i.e., a system of the form (III.3.1) for $f = R_2$, $g = R_1 - R_3$. The associated form (III.3.2) is

$$(2.16)\qquad \Phi = -2H(ar^2\mu^2 + 2brs\mu\nu + cs^2\nu^2),$$

and its discriminant is $\Delta = 4H^2Kr^2s^2 > 0$. Thus the system (2.15) is elliptic.

Now, see (1.18_2),

$$(2.17)\qquad \delta II = R_1(\omega^1)^2 + 2R_2\omega^1\omega^2 + R_3(\omega^2)^2,$$

and the condition (ii) implies $R_1 = R_3 = \lambda$, $R_2 = 0$ on ∂M. According to Theorem III.3.1, $R_1 - R_3 = R_2 = 0$ on M; $R_1 = R_3 = 0$ on M is then a consequence of (2.9). Thus $\delta II = 0$ on M. QED.

Let us present another version of the same theorem.

Theorem 2.2. *Let $M: \mathscr{D} \to E^3$, $\mathscr{D} \subset \mathbf{R}^2$ a bounded domain, be a surface. To M, associate a vector field v, and let Φ be the infinitesimal deformation of M induced by v (as described in the previous paragraph). Let us suppose:* (i) *Φ is an infinitesimal I-isometry*; (ii) *$K > 0$ on M*; (iii) *the vector $v(m)$ is orthogonal to $T_m(M)$ for each $m \in \partial M$. Then M is infinitesimally rigid with respect to Φ.*

Proof. Let us choose a field of frames associated to M, and let v be given by (1.7). Then, see (1.18) and $(1.16_{1,2})$,

$$(2.18)\qquad \frac{1}{2}\,\delta I = (x_1 - az)(\omega^1)^2 + (x_2 + y_1 - 2bz)\,\omega^1\omega^2 + (y_2 - cz)(\omega^2)^2$$

and (i) implies

$$(2.19)\qquad x_1 = az,\quad x_2 + y_1 = 2bz,\quad y_2 = cz.$$

The function p be defined by

$$(2.20)\qquad p = x_2 - bz = bz - y_1.$$

The equations $(1.8_{1,2})$ take the form

$$(2.21)\qquad \mathrm{d}x - y\omega_1{}^2 = az\omega^1 + (bz + p)\,\omega^2,$$

$$\mathrm{d}y + x\omega_1{}^2 = (bz - p)\,\omega^1 + cz\omega^2.$$

On M, let us choose isothermic coordinates and the frames in such a way that

$$(2.22)\qquad \omega^1 = r\,\mathrm{d}u,\quad \omega^2 = r\,\mathrm{d}v;\quad \omega_1{}^2 = -\frac{1}{r}\frac{\partial r}{\partial v}\mathrm{d}u + \frac{1}{r}\frac{\partial r}{\partial u}\mathrm{d}v.$$

The equations (2.21) take the form

$$(2.23)\qquad \frac{\partial x}{\partial u} + \frac{1}{r}\frac{\partial r}{\partial v}\,y = azr, \qquad \frac{\partial x}{\partial v} - \frac{1}{r}\frac{\partial r}{\partial u}\,y = (bz + p)\,r,$$

$$\frac{\partial y}{\partial u} - \frac{1}{r}\frac{\partial r}{\partial v}\,x = (bz - p)\,r, \qquad \frac{\partial y}{\partial v} + \frac{1}{r}\frac{\partial r}{\partial u}\,x = czr.$$

Multiplying (2.23) successively by $2b$, $-(a + c)$, $-(a + c)$, $2b$, and adding them together, we get

$$(2.24)\qquad 2b\left(\frac{\partial x}{\partial u} + \frac{\partial y}{\partial v}\right) - (a + c)\left(\frac{\partial x}{\partial v} + \frac{\partial y}{\partial u}\right) = (.)\,x + (.)\,y;$$

from $(2.23_{1,4})$,

$$(2.25)\qquad c\,\frac{\partial x}{\partial u} - a\,\frac{\partial y}{\partial v} = (.)\,x + (.)\,y.$$

The associated form of the system (2.24) + (2.25)

$$(2.26)\qquad \Phi = (a + c)\,(a\mu^2 + 2b\mu\nu + c\nu^2)$$

has the discriminant $\Delta = 4H^2K > 0$, and the considered system is elliptic. Because of (iii), $x = y = 0$ on ∂M, and Theorem III.3.1 implies $x = y = 0$ on M. From (2.21), $az = bz = cz = 0$, and this implies $z = 0$, i.e., $v = 0$ on M. QED.

The theorem on the rigidity being one of the most famous classical results, it is almost our duty to present another proof of it. This time we are going to make use of an integral formula; the idea of the proof goes back to W. Blaschke.

Theorem 2.3. *Let $M: \mathscr{D} \to E^3$ be a surface and Φ its infinitesimal I-isometry. Suppose:* (i) $K > 0$ *on* M; (ii) *there is* $\delta II = 0$ *on* ∂M. *Then* M *is infinitesimally rigid with respect to* Φ.

Proof. Let $M_t: \mathscr{D} \to E^3$ be a layer of surfaces creating Φ (as described in the previous paragraph); further, let $S \in E^3$ be a fixed point. For each $t \in I \subset \mathbf{R}$, consider the vector field v_t defined by

$$(2.27)\qquad S = m(t) + v_t,$$

let us write

$$(2.28)\qquad v_t = (q + tq' + \cdots)\,v_1(t) + (r + tr' + \cdots)\,v_2(t) + (p + tp' + \cdots)\,v_3(t).$$

Φ being an infinitesimal I-isometry, we are in the position to choose the frames of M_t in such a way that (2.5) and (2.6) hold true; see the beginning of the proof of Theorem 2.1. Thus the differential consequence of (2.27) is

$$\begin{aligned}0 = {} & \omega^1 v_1(t) + \omega^2 v_2(t) + (\mathrm{d}q + t\cdot\mathrm{d}q' + \cdots)\,v_1(t) + (dr + t\cdot\mathrm{d}r' + \cdots)\,v_2(t)\\ & + (\mathrm{d}p + t\cdot\mathrm{d}p' + \cdots)\,v_3(t)\\ & + (q + tq' + \cdots)\,\{\omega_1{}^2 v_2(t) + (\omega_1{}^3 + t\varphi_1{}^3 + \cdots)\,v^3(t)\}\\ & + (r + tr' + \cdots)\,\{-\omega_1{}^2 v_1(t) + (\omega_2{}^3 + t\varphi_2{}^3 + \cdots)\,v_3(t)\}\\ & - (p + tp' + \cdots)\,\{(\omega_1{}^3 + t\varphi_1{}^3 + \cdots)\,v_1(t) + (\omega_2{}^3 + t\varphi_2{}^3 + \cdots)\,v_2(t)\},\end{aligned}$$

i.e.,

$$\begin{aligned}(2.29)\qquad dq &= r\omega_1{}^2 + p\omega_1{}^3 - \omega^1,\\ dr &= -q\omega_1{}^2 + p\omega_2{}^3 - \omega^2,\\ dp &= -q\omega_1{}^3 - r\omega_2{}^3;\end{aligned}$$

$$\begin{aligned}(2.30)\qquad dq' &= r'\omega_1{}^2 + p\varphi_1{}^3 + p'\omega_1{}^3,\\ dr' &= -q'\omega_1{}^2 + p\varphi_2{}^3 + p'\omega_2{}^3,\\ dp' &= -q\varphi_1{}^3 - q'\omega_1{}^3 - r\varphi_2{}^3 - r'\omega_2{}^3.\end{aligned}$$

Recall that (2.5), (2.6) reduce (1.3)—(1.6) to

$$\begin{aligned}(2.31)\qquad &\omega^1 \wedge \varphi_1{}^3 + \omega^2 \wedge \varphi_2{}^3 = 0, \quad \omega_1{}^3 \wedge \varphi_2{}^3 + \varphi_1{}^3 \wedge \omega_2{}^3 = 0,\\ &d\varphi^1 = -\varphi^2 \wedge \omega_1{}^3, \quad d\varphi^2 = \varphi^1 \wedge \omega_1{}^2, \quad d\varphi_1{}^3 = \omega_1{}^2 \wedge \varphi_2{}^3, \quad d\varphi_2{}^3 = -\omega_1{}^2 \wedge \varphi_1{}^3.\end{aligned}$$

Let us change the frames of each M_t according to (1.21). Then

$$\begin{aligned}S &= m + ({}^*q + t\cdot{}^*q' + \cdots)\,{}^*v_1(t) + ({}^*r + t\cdot{}^*r' + \cdots)\,{}^*v_2(t)\\ &\quad + ({}^*p + t\cdot{}^*p' + \cdots)\,{}^*v_3(t)\\ &= m + (q + tq' + \cdots)\,\{\varepsilon_1(\cos\varrho - t\sigma\sin\varrho + \cdots)\,{}^*v_1(t)\\ &\quad - \varepsilon_1(\sin\varrho + t\sigma\cos\varrho + \cdots)\,{}^*v_2(t)\}\\ &\quad + (r + tr' + \cdots)\,\{(\sin\varrho + t\sigma\cos\varrho + \cdots)\,{}^*v_1(t)\\ &\quad + (\cos\varrho - t\sigma\sin\varrho + \cdots)\,{}^*v_2(t)\} + (p + tp' + \cdots)\,\varepsilon_2{}^*v_3(t),\end{aligned}$$

i.e.,

$$\begin{aligned}(2.32)\qquad &{}^*p = \varepsilon_2 p, \quad {}^*q = \varepsilon_1\cos\varrho\cdot q + \sin\varrho\cdot r, \quad {}^*r = -\varepsilon_1\sin\varrho\cdot q + \cos\varrho\cdot r;\\ &{}^*p' = \varepsilon_2 p', \quad {}^*q' = -\varepsilon_1\sigma\sin\varrho\cdot q + \sigma\cos\varrho\cdot r + \varepsilon_1\cos\varrho\cdot q' + \sin\varrho\cdot r',\\ &{}^*r' = -\varepsilon_1\sigma\cos\varrho\cdot q - \sigma\sin\varrho\cdot r - \varepsilon_1\sin\varrho\cdot q' + \cos\varrho\cdot r'.\end{aligned}$$

The frames have been chosen in such a way that $\varphi^1 = \varphi^2 = 0$; preserving this property, we should have ${}^*\varphi^1 = {}^*\varphi^2 = 0$, $(1.24_{1,2})$ implying

$$(2.33)\qquad \sigma = 0.$$

Now, let us consider the 1-form

$$(2.34)\qquad \tau = r'\varphi_1{}^3 - q'\varphi_2{}^3.$$

Then it is easy to see that

$$(2.35)\qquad {}^*\tau = \varepsilon_1\varepsilon_2\tau$$

and

$$(2.36)\qquad d\tau = -2p\varphi_1{}^3 \wedge \varphi_2{}^3$$

as a consequence of (2.30) and (2.31). From (2.31_1), we get the existence of functions R_1, R_2, R_3 such that

$$(2.37)\qquad \varphi_1{}^3 = R_1\omega^1 + R_2\omega^2, \qquad \varphi_2{}^3 = R_2\omega^1 + R_3\omega^2,$$

the conditions (2.31_2) turns out to be

(2.38) $\quad aR_3 - 2bR_2 + cR_1 = 0.$

This is equivalent to

(2.39) $\quad 2HR_1 = 2bR_2 - a(R_3 - R_1) \quad \text{or} \quad 2HR_3 = 2bR_2 + c(R_3 - R_1)$ resp.

From this,

(2.40)
$$4H^2(R_1R_3 - R_2^2) = \{4b^2 - (a + c)^2\} R_2^2 - 2b(a - c) R_2(R_3 - R_1) - ac(R_3 - R_1)^2.$$

The supposition (i) implies that the right-hand side form (regarded as a quadratic form in R_2 and $R_3 - R_1$) is negative definite, its discriminant being $4H^2K > 0$. Thus $K > 0$ implies

(2.41) $\quad R_1R_3 - R_2^2 \geqq 0.$

Because of (ii), $\varphi_1^3 = \varphi_2^3 = 0$ on ∂M, and we get the integral formula

(2.42)
$$0 = \int_M p(R_1R_3 - R_2^2)\, \omega^1 \wedge \omega^2.$$

Now, we shall state, without proof, a geometric consequence of (i): the point $S \in E^3$ may be chosen in such a way that $S \notin T_m(M)$ for each $m \in M$. This consequence follows from a more general result saying that $K > 0$ implies that M is a part of a boundary of a convex body in E^3. Of course, we could immediately suppose this consequence instead of (i). Because of this,

(2.43) $\quad p \neq 0$

on M, and (2.42), (2.41) imply

(2.44) $\quad R_1R_3 - R_2^2 = 0.$

Because of the definiteness of (2.40), $R_2 = R_1 - R_3 = 0$, i.e., $R_1 = R_2 = R_3 = 0$. QED.

Let us now turn our attention to the finite I-isometries. Be given two surfaces $M, \tilde{M} : \mathscr{D} \to E^3$. Let $\{\tilde{m}, \tilde{v}_1, \tilde{v}_2, \tilde{v}_3\}$ be frames associated to $\tilde{M}$, and let us write

(2.45)
$$\begin{aligned}
d\tilde{m} &= (\omega^1 + \tau^1)\, \tilde{v}_1 + (\omega^2 + \tau^2)\, \tilde{v}_2,\\
d\tilde{v}_1 &= (\omega_1^2 + \tau_1^2)\, \tilde{v}_2 + (\omega_1^3 + \tau_1^3)\, \tilde{v}_3,\\
d\tilde{v}_2 &= -(\omega_1^2 + \tau_1^2)\, \tilde{v}_1 + (\omega_3^2 + \tau_2^3)\, \tilde{v}_3,\\
d\tilde{v}_3 &= -(\omega_1^3 + \tau_1^3)\, \tilde{v}_1 - (\omega_2^3 + \tau_2^3)\, \tilde{v}_2.
\end{aligned}$$

From the integrability conditions for (2.45),

(2.46)
$$\omega^1 \wedge \tau_1^3 + \omega^2 \wedge \tau_2^3 + \tau^1 \wedge \omega_1^3 + \tau^2 \wedge \omega_2^3 + \tau^1 \wedge \tau_1^3 + \tau^2 \wedge \tau_2^3 = 0,$$

(2.47)
$$\begin{aligned}
d\tau^1 &= -\omega^2 \wedge \tau_1^2 - \tau^2 \wedge \omega_1^2 - \tau^2 \wedge \tau_1^2,\\
d\tau^2 &= \omega^1 \wedge \tau_1^2 + \tau^1 \wedge \omega_1^2 + \tau^1 \wedge \tau_1^2,
\end{aligned}$$

$$(2.48)\qquad d\tau_1{}^2 = -\omega_1{}^3 \wedge \tau_2{}^3 - \tau_1{}^3 \wedge \omega_2{}^3 - \tau_1{}^3 \wedge \tau_2{}^3,$$

$$(2.49)\qquad d\tau_1{}^3 = \omega_1{}^2 \wedge \tau_2{}^3 + \tau_1{}^2 \wedge \omega_2{}^3 + \tau_1{}^2 \wedge \tau_2{}^3,$$

$$d\tau_2{}^3 = -\omega_1{}^2 \wedge \tau_1{}^3 - \tau_1{}^2 \wedge \omega_1{}^3 - \tau_1{}^2 \wedge \tau_1{}^3.$$

Of course,

$$(2.50)\qquad \tilde{I} - I = 2(\omega^1\tau^1 + \omega^2\tau^2) + (\tau^1)^2 + (\tau^2)^2,$$

$$\widetilde{II} - II = \omega^1\tau_1{}^3 + \omega^2\tau_2{}^3 + \tau^1\omega_1{}^3 + \tau^2\omega_2{}^3 + \tau^1\tau_1{}^3 + \tau^2\tau_2{}^3,$$

$$\widetilde{III} - III = 2(\omega_1{}^3\tau_1{}^3 + \omega_2{}^3\tau_2{}^3) + (\tau_1{}^3)^2 + (\tau_2{}^3)^2.$$

Let

$$(2.51)\qquad \tilde{I} = I.$$

Then we are in the position to choose the frames $\{\tilde{m}, \tilde{v}_1, \tilde{v}_2, \tilde{v}_3\}$ in such a way that

$$(2.52)\qquad (d\tilde{M})^{-1}\,\tilde{v}_1 = (dM)^{-1}\,v_1, \qquad (d\tilde{M})^{-1}\,\tilde{v}_2 = (dM)^{-1}\,v_2,$$

and we get

$$(2.53)\qquad \tau^1 = \tau^2 = 0.$$

From (2.47),

$$(2.54)\qquad \tau_1{}^2 = 0,$$

and (2.46)—(2.49) reduce to

$$(2.55)\qquad \omega^1 \wedge \tau_1{}^3 + \omega^2 \wedge \tau_2{}^3 = 0, \quad \omega_1{}^3 \wedge \tau_2{}^3 + \tau_1{}^3 \wedge \omega_2{}^3 + \tau_1{}^3 \wedge \tau_2{}^3 = 0,$$

$$(2.56)\qquad d\tau_1{}^3 = \omega_1{}^2 \wedge \tau_2{}^3, \qquad d\tau_2{}^3 = -\omega_1{}^2 \wedge \tau_1{}^3.$$

Theorem 2.4. *Be given two I-isometric surfaces $M, \tilde{M} : \mathscr{D} \to E^3$, $\mathscr{D} \subset \mathbf{R}^2$ a bounded domain. Suppose:* (i) *$K > 0$ on M*; (ii) *$II = \widetilde{II}$ on ∂M. Then $II = \widetilde{II}$ on M, i.e., both the surfaces are equal up to a motion* (see Theorem IV.2.2).

Proof. From (2.55), we get the existence of functions $R_1, R_2, R_3 : \mathscr{D} \to \mathbf{R}$ such that

$$(2.57)\qquad \tau_1{}^3 = R_1\omega^1 + R_2\omega^2, \qquad \tau_2{}^3 = R_2\omega^1 + R_3\omega^2,$$

$$(2.58)\qquad cR_1 - 2bR_2 + aR_3 + R_1R_3 - R_2{}^2 = 0.$$

From (2.57), we get the existence of functions $T_1, \ldots, T_4 : \mathscr{D} \to \mathbf{R}$ satisfying

$$(2.59)\qquad dR_1 - 2R_2\omega_1{}^2 = T_1\omega^1 + T_2\omega^2,$$

$$dR_2 + (R_1 - R_3)\,\omega_1{}^2 = T_2\omega^1 + T_3\omega^2,$$

$$dR_3 + 2R_2\omega_1{}^2 = T_3\omega^1 + T_4\omega^2;$$

our starting point will consist of the equations

$$(2.60)\qquad d(R_1 + R_3) - 4R_2\omega_1{}^2 = (T_1 - T_3)\,\omega^1 + (T_2 - T_4)\,\omega^2,$$

$$dR_2 + (R_1 - R_3)\,\omega_1{}^2 = T_2\omega^1 + T_3\omega^2.$$

The equation (2.58) may be rewritten as

$$(2.61)\qquad (2c + R_3)\, R_1 - 2(2b + R_2)\, R_2 + (2a + R_1)\, R_3 = 0.$$

From

$$(2.62)\qquad 2\tilde{H}\omega^1 \wedge \omega^2 = (\omega_1{}^3 + \tau_1{}^3) \wedge \omega^2 + \omega^1 \wedge (\omega_2{}^3 + \tau_2{}^3),$$

$$(2.63)\qquad 2\tilde{H} = 2H + R_1 + R_3.$$

Because of $K = \tilde{K} > 0$, see supposition (i) and Theorem IV.2.1, we have $H\tilde{H} \neq 0$ on M. Thus (2.61) implies

$$(2.64)\qquad R_1 = \frac{1}{2\tilde{H}} \{(2a + R_1)(R_1 - R_3) + 2(2b + R_2)\, R_2\},$$

$$R_3 = \frac{1}{2\tilde{H}} \{-(2c + R_3)(R_1 - R_3) + 2(2b + R_2)\, R_2\}.$$

The differentiation of (2.58) implies

$$(2.65)\qquad cT_1 - 2bT_2 + aT_3 + (\gamma + T_3)\, R_1 - 2(\beta + T_2)\, R_2 + (\alpha + T_1)\, R_3 = 0,$$

$$cT_2 - 2bT_3 + aT_4 + (\delta + T_4)\, R_1 - 2(\gamma + T_3)\, R_2 + (\beta + T_2)\, R_3 = 0,$$

which we rewrite as

$$(2.66)\qquad cT_1 - 2bT_2 + aT_3 = (.)\,(R_1 - R_3) + (.)\, R_2,$$

$$cT_2 - 2bT_3 + aT_4 = (.)\,(R_1 - R_3) + (.)\, R_2$$

in accord with (2.64).

On $\mathscr{D}$, let us choose isothermic coordinates (u, v); the frames be chosen in such a way that we have (V.1.4) and (V.1.5) with $r = s$. Then (2.60) implies

$$(2.67)\qquad \frac{\partial(R_1 - R_3)}{\partial u} = (T_1 - T_3)\, r + (.)\,(R_1 - R_3) + (.)\, R_2,$$

$$\frac{\partial(R_1 - R_3)}{\partial v} = (T_2 - T_4)\, r + (.)\,(R_1 - R_3) + (.)\, R_2,$$

$$\frac{\partial R_2}{\partial u} = T_2 r + (.)\,(R_1 - R_3) + (.)\, R_2, \quad \frac{\partial R_2}{\partial v} = T_3 r + (.)\,(R_1 - R_3) + (.)\, R_2.$$

Calculating $T_1 r, \ldots, T_4 r$ from (2.67) and inserting them into (2.66), we get the system

$$(2.68)\qquad c\,\frac{\partial(R_1 - R_3)}{\partial u} - 2b\,\frac{\partial R_2}{\partial u} + (a + c)\,\frac{\partial R_2}{\partial v} = (.)\,(R_1 - R_3) + (.)\, R_2,$$

$$-a\,\frac{\partial(R_1 - R_3)}{\partial v} + (a + c)\,\frac{\partial R_2}{\partial u} - 2b\,\frac{\partial R_2}{\partial v} = (.)\,(R_1 - R_3) + (.)\, R_2.$$

The associated form being

$$(2.69)\qquad \Phi = (a + c)\,(a\mu^2 + 2b\mu\nu + c\nu^2),$$

the system (2.68) is elliptic because of $H^2K > 0$. From (2.50_2),

$$\widetilde{II} - II = R_1(\omega^1)^2 + 2R_2\omega^1\omega^2 + R_3(\omega^2)^2; \tag{2.70}$$

thus the supposition (ii) implies $R_1 = R_2 = R_3 = 0$ on ∂M. The ellipticity of (2.68) yields $R_1 - R_3 = R_2 = 0$ on M, from (2.64), $R_1 = R_3 = 0$. Thus $\widetilde{II} = II$ on M. QED.

It is easy to see that (ii) might be weakened to (ii'): there is a function $\lambda\colon \partial M \to \mathbf{R}$ such that $\widetilde{II} = II + \lambda \cdot I$ on ∂M.

Let us present another proof of Theorem 2.4 by means of an integral formula. We do not enter into the details; let us remark that we may remove the condition of the boundedness of $\mathscr{D}$.

Let $S, \tilde{S} \in E^3$ be fixed points; the vector fields v, $\tilde{v}$ be defined by

$$S = m + v, \quad \tilde{S} = \tilde{m} + \tilde{v}; \qquad v = qv_1 + rv_2 + pv_3, \quad \tilde{v} = \tilde{q}\tilde{v}_1 + \tilde{r}\tilde{v}_2 + \tilde{p}\tilde{v}_3. \tag{2.71}$$

Then — see (2.29) —

$$\begin{aligned} &dq - r\omega_1{}^2 - p\omega_1{}^3 + \omega^1 = 0, \quad dr + q\omega_1{}^2 - p\omega_2{}^3 + \omega^2 = 0, \\ &dp + q\omega_1{}^3 + r\omega_2{}^3 = 0; \\ &d\tilde{q} - \tilde{r}\omega_1{}^2 - \tilde{p}(\omega_1{}^3 + \tau_1{}^3) + \omega^1 = 0, \quad d\tilde{r} + \tilde{q}\omega_1{}^2 - \tilde{p}(\omega_2{}^3 + \tau_2{}^3) + \omega^2 = 0, \\ &d\tilde{p} + \tilde{q}(\omega_1{}^3 + \tau_1{}^3) + \tilde{r}(\omega_2{}^3 + \tau_2{}^3) = 0. \end{aligned} \tag{2.72}$$

Consider the 1-form

$$\tau = (r - \tilde{r})\,\tau_1{}^3 - (q - \tilde{q})\,\tau_2{}^3; \tag{2.73}$$

it may be proved that τ is invariant (up to a sign). Because of (ii), $\tau = 0$ on ∂M. We have

$$d\tau = 2\tilde{p}\tau_1{}^3 \wedge \tau_2{}^3, \tag{2.74}$$

and the integral formula

$$\int_M \tilde{p}(R_1R_3 - R_2{}^2)\,\omega^1 \wedge \omega^2 = 0; \tag{2.75}$$

following the ideas of the proof of Theorem 2.3, we might prove $R_1 = R_2 = R_3 = 0$.

3. *II*-isometries

I have to admit that our "technology" of ω-calculus does not seem (at least to me) convenient for handling the theory of finite *II*-isometries. Because of that I am going to restrict myself to the infinitesimal case. The theory of *II*-isometries is nowadays in the center of attention: the reasonable conjecture says that two *II*-isometric surfaces are *I*-isometric (supposing, for example, the positiveness of Gauss' curvatures). Nobody has been able to prove (or disprove) it; to ensure the *I*-isometry, the authors add other suppositions — mainly the conservation of certain combinations of curvatures. In what follows, I replace this by a more general inequality (see Theorem 3.3).

Let us prepare the common analytical backround for the proofs of our theorems. Write

$$(3.1)\qquad \varphi^1 = a_1\omega^1 + a_2\omega^2, \qquad \varphi^2 = a_3\omega^1 + a_4\omega^2,$$

$$(3.2)\qquad \varphi_1{}^3 = b_1\omega^1 + b_2\omega^2, \qquad \varphi_2{}^3 = b_3\omega^1 + b_4\omega^2;$$

(1.3) is then equal to

$$(3.3)\qquad b_2 - b_3 + ba_1 - aa_2 + ca_3 - ba_4 = 0,$$

and we have

$$(3.4)\qquad \frac{1}{2}\,\delta I = a_1(\omega^1)^2 + (a_2 + a_3)\,\omega^1\omega^2 + a_4(\omega^2)^2,$$

$$\delta II = (b_1 + aa_1 + ba_3)\,(\omega^1)^2 + (b_4 + ba_2 + ca_4)\,(\omega^2)^2 + (b_2 + b_3 + ba_1 + aa_2 + ca_3 + ba_4)\,\omega^1\omega^2.$$

Φ being an infinitesimal II-isometry, we have

$$(3.5)\qquad b_1 + aa_1 + ba_3 = 0, \quad b_2 + b_3 + ba_1 + aa_2 + ca_3 + ba_4 = 0, \quad b_4 + ba_2 + ca_4 = 0;$$

this together with (3.3) implies that the equations (3.2) read

$$(3.6)\qquad \varphi_1{}^3 = -a_1\omega_1{}^3 - a_3\omega_2{}^3, \qquad \varphi_2{}^3 = -a_2\omega_1{}^3 - a_4\omega_2{}^3.$$

From (3.1) and (3.6),

$$(3.7)\qquad \{da_1 - (a_2 + a_3)\,\omega_1{}^2\} \wedge \omega^1 + \{da_2 + (a_1 - a_4)\,\omega_1{}^2 - \varphi_1{}^2\} \wedge \omega^2 = 0,$$

$$\{da_3 + (a_1 - a_4)\,\omega_1{}^2 + \varphi_1{}^2\} \wedge \omega^1 + \{da_4 + (a_2 + a_3)\,\omega_1{}^2\} \wedge \omega^2 = 0,$$

$$\{da_1 - (a_2 + a_3)\,\omega_1{}^2\} \wedge \omega_1{}^3 + \{da_3 + (a_1 - a_4)\,\omega_1{}^2 + \varphi_1{}^2\} \wedge \omega_2{}^3 = 0,$$

$$\{da_2 + (a_1 - a_4)\,\omega_1{}^2 - \varphi_1{}^2\} \wedge \omega_1{}^3 + \{da_4 + (a_2 + a_3)\,\omega_1{}^2\} \wedge \omega_2{}^3 = 0,$$

and we get the existence of functions $A_1, \ldots, B_3$ such that

$$(3.8)\qquad da_1 - (a_2 + a_3)\,\omega_1{}^2 = A_1\omega^1 + A_2\omega^2,$$

$$da_2 + (a_1 - a_4)\,\omega_1{}^2 - \varphi_1{}^2 = A_2\omega^1 + A_3\omega^2,$$

$$da_3 + (a_1 - a_4)\,\omega_1{}^2 + \varphi_1{}^2 = B_1\omega^1 + B_2\omega^2,$$

$$da_4 + (a_2 + a_3)\,\omega_1{}^2 = B_2\omega^1 + B_3\omega^2,$$

$$(3.9)\qquad bA_1 - aA_2 + cB_1 - bB_2 = 0, \quad bA_2 - aA_3 + cB_2 - bB_3 = 0.$$

From (3.8),

$$(3.10)\qquad \{dA_1 - (2A_2 + B_1)\,\omega_1{}^2\} \wedge \omega^1 + \{dA_2 + (A_1 - A_3 - B_2)\,\omega_1{}^2\} \wedge \omega^2 = (a_2 + a_3)\,K\omega^1 \wedge \omega^2,$$

$$\{dA_2 + (A_1 - A_3 - B_2)\,\omega_1{}^2\} \wedge \omega^1 + \{dA_3 + (2A_2 - B_3)\,\omega_1{}^2\} \wedge \omega^2 = -2a_1K\omega^1 \wedge \omega^2,$$

$$\{dB_1 - (2B_2 - A_1)\,\omega_1{}^2\} \wedge \omega^1 + \{dB_2 + (B_1 - B_3 + A_2)\,\omega_1{}^2\} \wedge \omega^2$$
$$= 2a_4K\omega^1 \wedge \omega^2,$$
$$\{dB_2 + (B_1 - B_3 + A_2)\,\omega_1{}^2\} \wedge \omega^1 + \{dB_3 + (2B_2 + A_3)\,\omega_1{}^2\} \wedge \omega^2$$
$$= -(a_2 + a_3)\,K\omega^1 \wedge \omega^2,$$

and we get the existence of functions $C_1, \ldots, C_8$ such that

$$\begin{aligned} &dA_1 - (2A_2 + B_1)\,\omega_1{}^2 = C_1\omega^1 + (C_2 - a_2K)\,\omega^2, \\ &dA_2 + (A_1 - A_3 - B_2)\,\omega_1{}^2 = (C_2 + a_3K)\,\omega^1 + (C_3 + a_1K)\,\omega^2, \\ &dA_3 + (2A_2 - B_3)\,\omega_1{}^2 = (C_3 - a_1K)\,\omega^1 + C_4\omega^2, \\ &dB_1 - (2B_2 - A_1)\,\omega_1{}^2 = C_5\omega^1 + (C_6 - a_4K)\,\omega^2, \\ &dB_2 + (B_1 - B_3 + A_2)\,\omega_1{}^2 = (C_6 + a_4K)\,\omega^1 + (C_7 + a_2K)\,\omega^2, \\ &dB_3 + (2B_2 + A_3)\,\omega_1{}^2 = (C_7 - a_3K)\,\omega^1 + C_8\omega^2. \end{aligned} \tag{3.11}$$

The differential consequences of (3.9) are

$$\begin{aligned} &\beta A_1 - \alpha A_2 + \gamma B_1 - \beta B_2 + bC_1 - aC_2 + cC_5 - bC_6 - (aa_3 + ba_4)\,K = 0, \\ &\gamma A_1 - \beta A_2 + \delta B_1 - \gamma B_2 + bC_2 - aC_3 + cC_6 - bC_7 \\ &- (aa_1 + 2ba_2 + ca_4)\,K = 0, \\ &\beta A_2 - \alpha A_3 + \gamma B_2 - \beta B_3 + bC_2 - aC_3 + cC_6 - bC_7 \\ &+ (aa_1 + 2ba_3 + ca_4)\,K = 0, \\ &\gamma A_2 - \beta A_3 + \delta B_2 - \gamma B_3 + bC_3 - aC_4 + cC_7 - bC_8 + (ba_1 + ca_2)\,K = 0. \end{aligned} \tag{3.12}$$

We have

$$\delta K = -2K(a_1 + a_4), \quad \delta H = -(aa_1 + ba_2 + ba_3 + ca_4). \tag{3.13}$$

From (1.23) and (1.24),

$$\begin{aligned} {}^*a_1 &= \cos^2\varrho \cdot a_1 + \varepsilon_1 \sin\varrho\cos\varrho \cdot a_2 + \varepsilon_1 \sin\varrho\cos\varrho \cdot a_3 + \sin^2\varrho \cdot a_4, \\ {}^*a_2 &= \sigma - \sin\varrho\cos\varrho \cdot a_1 + \varepsilon_1 \cos^2\varrho \cdot a_2 - \varepsilon_1 \sin^2\varrho \cdot a_3 + \sin\varrho\cos\varrho \cdot a_4, \\ {}^*a_3 &= -\sigma - \sin\varrho\cos\varrho \cdot a_1 - \varepsilon_1 \sin^2\varrho \cdot a_2 + \varepsilon_1 \cos^2\varrho \cdot a_3 + \sin\varrho\cos\varrho \cdot a_4, \\ {}^*a_4 &= \sin^2\varrho \cdot a_1 - \varepsilon_1 \sin\varrho\cos\varrho \cdot a_2 - \varepsilon_1 \sin\varrho\cos\varrho \cdot a_3 + \cos^2\varrho \cdot a_4, \end{aligned} \tag{3.14}$$

this implying

$$\begin{aligned} {}^*a_1 - {}^*a_4 &= \cos 2\varrho \cdot (a_1 - a_4) + \varepsilon_1 \sin 2\varrho \cdot (a_2 + a_3), \\ {}^*a_2 + {}^*a_3 &= -\sin 2\varrho \cdot (a_1 - a_4) + \varepsilon_1 \cos 2\varrho \cdot (a_2 + a_3). \end{aligned} \tag{3.15}$$

We have

$$\begin{aligned} &d(a_1 - a_4) - 2(a_2 + a_3)\,\omega_1{}^2 = (A_1 - B_2)\,\omega^1 + (A_2 - B_3)\,\omega^2, \\ &d(a_2 + a_3) + 2(a_1 - a_4)\,\omega_1{}^2 = (A_2 + B_1)\,\omega^1 + (A_3 + B_2)\,\omega^2; \end{aligned} \tag{3.16}$$

this and (3.15) yields

$$\begin{aligned} {}^*A_1 - {}^*B_2 = {} &\varepsilon_1 \cos\varrho\cos 2\varrho \cdot (A_1 - B_2) + \cos\varrho\sin 2\varrho \cdot (A_2 + B_1) \\ &+ \sin\varrho\cos 2\varrho \cdot (A_2 - B_3) + \varepsilon_1 \sin\varrho\sin 2\varrho \cdot (A_3 + B_2), \end{aligned} \tag{3.17}$$

$$\begin{aligned}
{}^*A_2 - {}^*B_3 &= -\varepsilon_1 \sin\varrho \cos 2\varrho \cdot (A_1 - B_2) - \sin\varrho \sin 2\varrho \cdot (A_2 + B_1) \\
&\quad + \cos\varrho \cos 2\varrho \cdot (A_2 - B_3) + \varepsilon_1 \cos\varrho \sin 2\varrho \cdot (A_3 + B_2), \\
{}^*A_2 + {}^*B_1 &= -\varepsilon_1 \cos\varrho \sin 2\varrho \cdot (A_1 - B_2) + \cos\varrho \cos 2\varrho \cdot (A_2 + B_1) \\
&\quad - \sin\varrho \sin 2\varrho \cdot (A_2 - B_3) + \varepsilon_1 \sin\varrho \cos 2\varrho \cdot (A_3 + B_2), \\
{}^*A_3 + {}^*B_2 &= \varepsilon_1 \sin\varrho \sin 2\varrho \cdot (A_1 - B_2) - \sin\varrho \cos 2\varrho \cdot (A_2 + B_1) \\
&\quad - \cos\varrho \sin 2\varrho \cdot (A_2 - B_3) + \varepsilon_1 \cos\varrho \cos 2\varrho \cdot (A_3 + B_2).
\end{aligned}$$

From the preceding relations, we get

Lemma 3.1. *Let Φ be an infinitesimal II-isometry of M. Consider the function*

$$J_\Phi = (A_1 - B_2)(A_3 + B_2) - (A_2 - B_3)(A_2 + B_1) \tag{3.18}$$

and the 1-form

$$\begin{aligned}
\varkappa_\Phi &= \{(a_1 - a_4)(A_2 + B_1) - (a_2 + a_3)(A_1 - B_2)\}\,\omega^1 \\
&\quad + \{(a_1 - a_4)(A_3 + B_2) - (a_2 + a_3)(A_2 - B_3)\}\,\omega^2.
\end{aligned} \tag{3.19}$$

Then

$${}^*J_\Phi = J_\Phi \tag{3.20}$$

and

$${}^*\varkappa_\Phi = \varepsilon_1 \varkappa_\Phi. \tag{3.21}$$

Theorem 3.1. *Let $M\colon \mathscr{D} \to E^3$, $\mathscr{D} \subset \mathbf{R}^2$ a bounded domain, be a surface and Φ its infinitesimal II-isometry, and let us suppose:* (i) *there are functions $R, S\colon M \to \mathbf{R}$ such that*

$$R \cdot \delta K + S \cdot \delta H = 0, \tag{3.22}$$

$$H^2K^2(2KR + HS)^2(4KR^2 + 4HRS + S^2) > 0 \tag{3.23}$$

on M; (ii) *there is a function $\lambda\colon \partial M \to \mathbf{R}$ such that $\delta I = \lambda \cdot I$ on ∂M. Then M is infinitesimally rigid with respect to Φ.*

Proof. Because of (3.13), the relation (3.22) turns out to be

$$(2KR + aS)\,a_1 + bS(a_2 + a_3) + (2KR + cS)\,a_4 = 0; \tag{3.24}$$

because of $2KR + HS \neq 0$, (3.24) may be rewritten as

$$a_1 = \frac{1}{2(2KR + HS)}\{(2KR + cS)(a_1 - a_4) - bS(a_2 + a_3)\} \tag{3.25}$$

or

$$a_4 = \frac{-1}{2(2KR + HS)}\{(2KR + aS)(a_1 - a_4) + bS(a_2 + a_3)\}$$

resp. The differential consequences of (3.24) may be written in the form

$$\begin{aligned}
&(2KR + aS)\,A_1 + (2KR + cS)\,B_2 + bS(A_2 + B_1) \\
&= (.)\,(a_1 - a_4) + (.)\,(a_2 + a_3), \\
&(2KR + aS)\,A_2 + (2KR + cS)\,B_3 + bS(A_3 + B_2) \\
&= (.)\,(a_1 - a_4) + (.)\,(a_2 + a_3).
\end{aligned} \tag{3.26}$$

Multiplying (3.26_2) by H, (3.9_1) by $2KR + HS$ and adding them together, we get

$$(3.27)\quad \begin{aligned} &b(2KR + HS)(A_1 - B_2) - H(2KR + cS)(A_2 - B_3) \\ &+ c(2KR + HS)(A_2 + B_1) + bHS(A_3 + B_2) \\ &= (.)(a_1 - a_4) + (.)(a_2 + a_3). \end{aligned}$$

Similarly, multiplying (3.26_1) by H and (3.9_2) by $-(2KR + HS)$, we get

$$(3.28)\quad \begin{aligned} &H(2KR + aS)(A_1 - B_2) - b(2KR + HS)(A_2 - B_3) \\ &+ bHS(A_2 + B_1) + a(2KR + HS)(A_3 + B_2) \\ &= (.)(a_1 - a_4) + (.)(a_2 + a_3). \end{aligned}$$

Introducing, in $\mathscr{D}$, isometric coordinates and the frames in the already many times used way (i.e., $\omega^1 = r\,du$, $\omega^2 = r\,dv$), (3.16) take the form

$$(3.29)\quad \begin{aligned} &\frac{\partial(a_1 - a_4)}{\partial u} + (.)(a_2 + a_3) = (A_1 - B_2)\,r, \\ &\frac{\partial(a_1 - a_4)}{\partial v} + (.)(a_2 + a_3) = (A_2 - B_3)\,r, \\ &\frac{\partial(a_2 + a_3)}{\partial u} + (.)(a_1 - a_4) = (A_2 + B_1)\,r, \\ &\frac{\partial(a_2 + a_3)}{\partial v} + (.)(a_1 - a_4) = (A_3 + B_2)\,r. \end{aligned}$$

Inserting these into (3.27) + (3.28), we get the system of partial differential equations

$$(3.30)\quad \begin{aligned} &b(2KR + HS)\frac{\partial(a_1 - a_4)}{\partial u} - H(2KR + cS)\frac{\partial(a_1 - a_4)}{\partial v} \\ &+ c(2KR + HS)\frac{\partial(a_2 + a_3)}{\partial u} + bHS\frac{\partial(a_2 + a_3)}{\partial v} \\ &= (.)(a_1 - a_4) + (.)(a_2 + a_3), \\ &H(2KR + aS)\frac{\partial(a_1 - a_4)}{\partial u} - b(2KR + HS)\frac{\partial(a_1 - a_4)}{\partial v} \\ &+ bHS\frac{\partial(a_2 + a_3)}{\partial u} + a\,(2KR + HS)\frac{\partial(a_2 + a_3)}{\partial v} \\ &= (.)(a_1 - a_4) + (.)(a_2 + a_3). \end{aligned}$$

The associated form being

$$(3.31)\quad \Phi = -HK(2KR + HS)\{(2aR + S)\,\mu^2 + 4bR\mu\nu + (2cR + S)\,\nu^2\},$$

the system (3.30) is elliptic as a consequence of (3.23). From (ii) and (3.4), $a_1 = a_4 = \lambda$, $a_2 + a_3 = 0$ on ∂M, thus $a_1 - a_4 = a_2 + a_3 = 0$ on M. From (3.8), $A_1 = B_2$, $A_2 = B_3$, $A_2 + B_1 = A_3 + B_2 = 0$, i.e., $B_1 = -A_2$, $B_2 = A_1$, $B_3 = A_2$, $A_3 = -A_1$. The equations (3.9) reduce to $HA_2 = HA_1 = 0$, and we get $A_1 = A_2 = A_3 = B_1 = B_2 = B_3 = 0$ from $H \neq 0$. Because of ($3.12_{1,3}$), $HK(a_1 + a_4) = 0$, and this implies $a_1 = a_4 = 0$ on M. Thus $\delta I = 0$ on M. QED.

Theorem 3.2. *Let* $M: \mathscr{D} \to E^3$, $\mathscr{D} \subset \mathbf{R}^2$ *a bounded domain, be a surface and* Φ *its infinitesimal II-isometry satisfying:* (i) $K > 0$ *on* M; (ii) $\delta \log K = \text{const}$; (iii) *there is a function* $\lambda: \partial M \to \mathbf{R}$ *such that* $\delta I = \lambda \cdot I$ *on* ∂M. *Then* M *is infinitesimally rigid with respect to* Φ.

Proof. From (3.13_1), $\delta \log K = -2(a_1 + a_4) = \text{const}$. Thus

$$(3.32) \qquad A_1 + B_2 = 0, \quad A_2 + B_4 = 0.$$

The equations (3.9) reduce to

$$(3.33) \qquad 2bA_1 - aA_2 + cB_1 = 0, \quad cA_1 - 2bA_2 + aA_3 = 0;$$

inserting into them the corresponding consequences of the reduced equations (3.29), we get

$$(3.34) \qquad b\,\frac{\partial(a_1 - a_4)}{\partial u} - H\,\frac{\partial(a_1 - a_4)}{\partial v} + c\,\frac{\partial(a_2 + a_3)}{\partial u}$$
$$= (.)\,(a_1 - a_4) + (.)\,(a_2 + a_3),$$
$$H\,\frac{\partial(a_1 - a_4)}{\partial u} - b\,\frac{\partial(a_1 - a_4)}{\partial v} + a\,\frac{\partial(a_2 + a_3)}{\partial v}$$
$$= (.)\,(a_1 - a_4) + (.)\,(a_2 + a_3).$$

The associated form

$$(3.35) \qquad \Phi = -H(a\mu^2 + 2b\mu\nu + c\nu^2)$$

being definite, the system (3.34) is elliptic, and we follow the end of the proof of the previous theorem. QED.

The just proved theorems are just more or less obvious generalizations of known results. On the contrary, the next theorem is, at least in my opinion, quite new. For the definition of J_Φ, recall Lemma 3.1.

Theorem 3.3. *Let* $M: \mathscr{D} \to E^3$ *be a surface and* Φ *its infinitesimal II-isometry. Suppose:* (i) $K > 0$ *and* $J_\Phi \leqq 0$ *on* M; (ii) *there is a function* $\lambda: \partial M \to \mathbf{R}$ *such that* $\delta I = \lambda \cdot I$ *on* ∂M. *Then* M *is infinitesimally rigid with respect to* Φ. *The supposition* (i) *may be replaced by* (i'): $K < 0$, $H \neq 0$ *and* $J_\Phi \geqq 0$ *on* M.

Proof. The assertion follows immediately from the integral formula

$$(3.36) \qquad \int_{\partial M} \varkappa_\Phi = 2 \int_M \{J_\Phi - [(a_1 - a_4)^2 + (a_2 + a_3)^2]\, K\}\, do.$$

QED.

Two classic results are now easy consequences of this theorem.

Corollary 3.1 (K. Voss). *Let* M *be a part of a sphere and* Φ *be its infinitesimal II-isometry. Suppose that there is a function* $\lambda: \partial M \to \mathbf{R}$ *such that* $\delta I = \lambda \cdot I$ *on* ∂M. *Then* M *is infinitesimally rigid with respect to* Φ.

Proof. For the sphere M,

$$(3.37) \qquad 0 \neq a = c = \text{const}, \quad b = 0, \quad \alpha = \beta = \gamma = \delta = 0.$$

From (3.9), $A_2 = B_1$ and $A_3 = B_2$; from $(3.12_{2,3})$, $a_1 + a_4 = 0$. Finally, from $(3.8_{1,4})$, $A_1 + B_2 = A_2 + B_3 = 0$. Thus $A_3 = B_2 = -A_1$, $B_1 = -B_3 = A_3$ and

$$(3.38) \qquad J_\Phi = -4(A_1^2 + A_2^2) \leqq 0.$$

Now, apply Theorem 3.3. QED.

Corollary 3.2 (V. G. Grove). *Let $M: \mathscr{D} \to E^3$ be a surface and Φ its infinitesimal II-isometry. Suppose:* (i) $K > 0$ *on* M; (ii) $\delta K = 0$ *on* M; (iii) *there is a function* $\lambda: \partial M \to \mathbf{R}$ *such that* $\delta I = \lambda \cdot I$ *on* ∂M. *Then M is infinitesimally rigid with respect to Φ.*

Proof. From (3.13_1), $a_1 + a_4 = 0$, and (3.8) implies $A_1 + B_2 = A_2 + B_3 = 0$. Let $m \in M$ be a generic point; the frames of M be chosen in such a way that $b(m) = 0$ and $H > 0$. The equations (3.9) reduce to $aA_2 = cB_1$, $aA_3 = cB_2$ at m. Thus there are numbers p, q such that $A_2 = pc$, $B_1 = pa$, $A_3 = qc$, $B_2 = qa$, and we get

$$(3.39) \qquad J_\Phi(m) = -4H(aq^2 + cp^2) \leqq 0.$$

Now, apply Theorem 3.3. QED.

Of course, Corollary 3.2 is a consequence of Theorem 3.2 as well.

Using Theorems III.3.3—5, we are now in the position to prove more fine results generalizing the K. Voss' theorem (Corollary 3.1).

Thus, let M be a domain of a unit sphere and let Φ be its infinitesimal II-isometry. Then

$$(3.40) \qquad \omega_1^3 = \omega^1, \quad \omega_2^3 = \omega^2, \quad \text{i.e.,} \quad a = c = 1, \quad b = 0,$$

and (3.9) and (3.8) imply the existence of functions $A_1, \ldots, A_4$ such that

$$(3.41) \qquad \begin{aligned} &\mathrm{d}a_1 - (a_2 + a_3)\,\omega_1^2 = A_1\omega^1 + A_2\omega^2, \\ &\mathrm{d}a_2 + (a_1 - a_4)\,\omega_1^2 - \varphi_1^2 = A_2\omega^1 + A_3\omega^2, \\ &\mathrm{d}a_3 + (a_1 - a_4)\,\omega_1^2 + \varphi_1^2 = A_2\omega^1 + A_3\omega^2, \\ &\mathrm{d}a_4 + (a_2 + a_3)\,\omega_1^2 = A_3\omega^1 + A_4\omega^2. \end{aligned}$$

We get $\mathrm{d}(a_2 - a_3) - 2\varphi_1^2$ and

$$(3.42) \qquad 0 = \omega^1 \wedge \varphi_2^3 + \varphi_1^3 \wedge \omega^2 = -(a_1 + a_4)\,\omega^1 \wedge \omega^2, \quad \text{i.e.,} \quad a_1 + a_4 = 0.$$

Thus

$$(3.43) \qquad A_1 + A_3 = A_2 + A_4 = 0,$$

and our fundamental equations for Φ reduce to

$$(3.44) \qquad \begin{aligned} &\varphi^1 = a_1\omega^1 + a_2\omega^2, \quad \varphi^2 = a_3\omega^1 - a_1\omega^2, \quad \varphi_1^3 = -a_1\omega^1 - a_3\omega^2, \\ &\varphi_2^3 = -a_2\omega^1 + a_1\omega^2. \end{aligned}$$

On M, let us consider a system of isothermic coordinates and the frames in such a way that

$$(3.45) \qquad \omega^1 = r\,\mathrm{d}u, \quad \omega^2 = r\,\mathrm{d}v, \quad \omega_1^2 = \frac{1}{r}\left(-\frac{\partial r}{\partial v}\,\mathrm{d}u + \frac{\partial r}{\partial u}\,\mathrm{d}v\right).$$

From (3.41), we get

$$(3.46)\qquad \frac{\partial(a_2+a_3)}{\partial u} - 2\frac{\partial a_1}{\partial v} + \frac{2}{r}\frac{\partial r}{\partial u}(a_2+a_3) - \frac{4}{r}\frac{\partial r}{\partial v}a_1 = 0,$$

$$\frac{\partial(a_2+a_3)}{\partial v} + 2\frac{\partial a_1}{\partial u} + \frac{2}{r}\frac{\partial r}{\partial v}(a_2+a_3) + \frac{4}{r}\frac{\partial r}{\partial u}a_1 = 0,$$

i.e.,

$$(3.47)\qquad \frac{\partial}{\partial u}\{r^2(a_2+a_3)\} - \frac{\partial}{\partial v}(2r^2a_1) = 0,$$

$$\frac{\partial}{\partial v}\{r^2(a_2+a_3)\} + \frac{\partial}{\partial u}(2r^2a_1) = 0.$$

From (3.44),

$$(3.48)\qquad \delta I = 2r^2\{a_1\,du^2 + (a_2+a_3)\,du\,dv - a_1\,dv^2\}.$$

Lemma 3.2. *Let M be a domain of a unit sphere homeomorphic with a bounded domain of $\mathbf{R}^2$. Denote by $\mathscr{L}$ the vector space over $\mathbf{R}$ of the forms δI induced by infinitesimal II-isometries Φ of M. On M, let us choose a system of isothermic coordinates (u, v); let v_1 and v_2 be the unit tangent vector fields of the curves $v =$ const and $u =$ const resp. Let*

$$(3.49)\qquad \Omega = c_1\,du^2 + 2c_2\,du\,dv + c_3\,dv^2$$

be a quadratic differential form on M. Then $\Omega \in \mathscr{L}$ if and only if $\Omega(v_1) + \Omega(v_2) = 0$ and the function $r^2\Omega(v_1 - iv_2)$ is holomorphic.

Proof. From (3.47) follows that the function

$$(3.50)\qquad f = r^2(a_2 + a_3 + 2a_1 i) = \frac{1}{2}r^2 i\cdot\delta I(v_1 - iv_2)$$

is holomorphic. Conversely, let Ω (3.49) satisfy the conditions of our lemma. Then $c_3 = -c_1$. Take

$$(3.51)\qquad a_1 = \frac{c_1}{2r^2},\qquad a_2 = a_3 = \frac{c_2}{2r^2}.$$

It is easy to show that the forms (3.44) with (3.51) together with $\varphi_1{}^2 = 0$ are an infinitesimal II-isometry. QED.

The function

$$(3.52)\qquad w = a_2 + a_3 + 2a_1 i$$

satisfies, because of (3.46), the differential equation

$$(3.53)\qquad \frac{\partial w}{\partial \bar z} + \frac{2}{r}\frac{\partial r}{\partial \bar z}w = 0.$$

The preceding discussion makes it clear that the solutions of (3.53) are in a 1-1-correspondence with the elements of $\mathscr{L}$. It is also easy to see that the zeros of w correspond to the *stationary points* of Φ, this being the points with $\delta I = 0$.

Along the boundary ∂M, take a field of unit tangent vectors

$$v = xv_1 + yv_2; \tag{3.54}$$

we have

$$\delta I(v) = 2xy(a_2 + a_3) + 2(x^2 - y^2)\, a_1. \tag{3.55}$$

Consider the function $\lambda\colon \partial M \to \mathbf{C}$ defined by

$$\lambda = 2xy + (x^2 - y^2)\,\mathrm{i}. \tag{3.56}$$

It is easy to see that, on ∂M,

$$\operatorname{Re}\left[\overline{\lambda(z)}\, w(z)\right] = \delta I\big(v(z)\big), \qquad |\lambda(z)| = 1. \tag{3.57}$$

Denote by $\mathscr{L}_v \subset \mathscr{L}$ the subspace of forms δI induced by infinitesimal *II*-isometries of M and satisfying $\delta I(v) = 0$ on ∂M; the elements of $\mathscr{L}_v$ are obviously in a 1-1-correspondence with the solutions of the problem $\mathbf{A}^0$ given by (3.53) and (3.57), and we may apply the above mentioned theorems. One of the easy consequences is formulated in the following

Theorem 3.4. *Let M be a surface diffeomorphic to a bounded domain of $\mathbf{R}^2$, a part of a sphere and $(m + 1)$-connected. Along its boundary ∂M be given a field v of unit tangent vectors. Denote by $\mathscr{L}_v$ the space of forms δI generated by infinitesimal II-isometries of M satisfying $\delta I(v) = 0$. Suppose $m = 0$, $\dim \mathscr{L}_v \geqq 1$ or $m \geqq 1$, $\dim \mathscr{L}_v \geqq 2m$ resp. If Φ is a non-trivial infinitesimal II-isometry of M with the corresponding $\delta I \in \mathscr{L}_v$, then $S_{\partial M} + 2S_M \leqq \dim \mathscr{L}_v + m - 1$, $S_{\partial m}$ and S_M being the number of the stationary points of Φ on the boundary ∂M or inside of M resp.*

4. *III*-isometries

We are going to prove just one result.

Theorem 4.1. *Let $M\colon \mathscr{D} \to E^3$, $\mathscr{D} \subset \mathbf{R}^2$ a bounded domain, be a surface and Φ its infinitesimal III-isometry. Suppose:* (i) *$K > 0$ on M;* (ii) *on M, there are functions P, Q such that*

$$P \cdot \delta\left(\frac{1}{K}\right) + 2Q \cdot \delta\left(\frac{H}{K}\right) = 0, \quad P^2 + 2HPQ + KQ^2 > 0; \tag{4.1}$$

(iii) *there is a function $\lambda\colon \partial M \to \mathbf{R}$ such that $\delta II = \lambda \cdot III$. Then M is infinitesimally rigid with respect to Φ.*

Proof. Because of $\omega_1^3 \wedge \omega_2^3 = K\omega^1 \wedge \omega^2$ and $K > 0$, the 1-forms ω_1^3, ω_2^3 are linearly independent. Thus the equation (IV.2.2) may be solved as well in the form

$$\omega^1 = a_1\omega_1^3 + b_1\omega_2^3, \qquad \omega^2 = b_1\omega_1^3 + c_1\omega_2^3, \tag{4.2}$$

the prolongation being

$$\begin{aligned} \mathrm{d}a_1 - 2b_1\omega_1^2 &= \alpha_1\omega_1^3 + \beta_1\omega_2^3, \\ \mathrm{d}b_1 + (a_1 - c_1)\,\omega_1^2 &= \beta_1\omega_1^3 + \gamma_1\omega_2^3, \\ \mathrm{d}c_1 + 2b_1\omega_1^2 &= \gamma_1\omega_1^3 + \delta_1\omega_2^3. \end{aligned} \tag{4.3}$$

With our experience, we may continue without delay. We may suppose

$$\varphi_1^{\ 3} = \varphi_2^{\ 3} = 0, \tag{4.4}$$

this implying

$$\varphi_1^{\ 2} = 0. \tag{4.5}$$

From (1.3),

$$\varphi^1 = R_1\omega_1^{\ 3} + R_2\omega_2^{\ 3}, \qquad \varphi^2 = R_2\omega_1^{\ 3} + R_3\omega_2^{\ 3} \tag{4.6}$$

and

$$\begin{aligned} &\mathrm{d}R_1 - 2R_2\omega_1^{\ 2} = S_1\omega_1^{\ 3} + S_2\omega_2^{\ 3}, \\ &\mathrm{d}R_2 + (R_1 - R_3)\,\omega_1^{\ 2} = S_2\omega_1^{\ 3} + S_3\omega_2^{\ 3}, \\ &\mathrm{d}R_3 + 2R_2\omega_1^{\ 2} = S_3\omega_1^{\ 3} + S_4\omega_2^{\ 3}. \end{aligned} \tag{4.7}$$

We have

$$\begin{aligned} &K = \frac{1}{a_1c_1 - b_1^{\ 2}}, \quad 2H = \frac{a_1 + c_1}{a_1c_1 - b_1^{\ 2}}, \\ &\delta K = -K^2(c_1R_1 - 2b_1R_2 + a_1R_3), \\ &2\delta H = K(R_1 + R_3) - 2HK(c_1R_1 - 2b_1R_2 + a_1R_3), \\ &\delta\left(\frac{1}{K}\right) = c_1R_1 - 2b_1R_2 - a_1R_3, \quad 2\delta\left(\frac{H}{K}\right) = R_1 + R_3 \end{aligned} \tag{4.8}$$

and (4.1_1) takes the form

$$(Pc_1 + Q)\,R_1 - 2Pb_1R_2 + (Pa_1 + Q)\,R_3 = 0. \tag{4.9}$$

The inequality (4.1_2) implies

$$HP + KQ \neq 0; \tag{4.10}$$

indeed, $HP + KQ = 0$ would make the left-hand side of (4.2_2) equal to $P^2K^{-1}(K - H^2) \leq 0$. We easily see that (4.9) and (4.10) make possible to express R_1 and R_3 as linear combinations of R_2 and $R_1 - R_3$. The differential consequences of (4.9) are

$$\begin{aligned} &(Pc_1 + Q)\,S_1 - 2Pb_1S_2 + (Pa_1 + Q)\,S_3 = (.)\,(R_1 - R_3) + (.)\,R_2, \\ &(Pc_1 + Q)\,S_2 - 2Pb_1S_3 + (Pa_1 + Q)\,S_4 = (.)\,(R_1 - R_3) + (.)\,R_2. \end{aligned} \tag{4.11}$$

On M, let us introduce coordinates (u, v) such that

$$\omega_1^{\ 3} = r\,\mathrm{d}u, \quad \omega_2^{\ 3} = s\,\mathrm{d}v, \quad rs \neq 0. \tag{4.12}$$

Then $\omega_1^{\ 2}$ is a linear combination of $\mathrm{d}u$, $\mathrm{d}v$, and (4.7) imply that (4.11) may be written as

$$\begin{aligned} &(Pc_1 + Q)\,s\,\frac{\partial(R_1 - R_3)}{\partial u} - 2Pb_1s\,\frac{\partial R_2}{\partial u} + (Pa_1 + Pc_1 + 2Q)\,r\,\frac{\partial R_2}{\partial v} \\ &\qquad = (.)\,(R_1 - R_3) + (.)\,R_2, \end{aligned} \tag{4.13}$$

$$- (Pa_1 + Q)\, r \,\frac{\partial(R_1 - R_3)}{\partial v} + (Pa_1 + Pc_1 + 2Q)\, s \,\frac{\partial R_2}{\partial u} - 2Pb_1 r \,\frac{\partial R_2}{\partial v}$$

$$= (.)\,(R_1 - R_3) + (.)\, R_2 .$$

The associated form is equal to

$$\Phi = (Pa_1 + Pc_1 + 2Q)\,\{(Pa_1 + Q)\, r^2\mu^2 + 2Pb_1 rs\mu\nu + (Pc_1 + Q)\, s^2\nu^2\}, \tag{4.14}$$

its discriminant to

$$\Delta = 4K^{-3}(PH + QK)^2\,(P^2 + 2HPQ + KQ^2)\, r^2 s^2 > 0 . \tag{4.15}$$

The rest of the proof is standard. QED.

VII. Surfaces in E^4 and E^5

1. Surfaces in E^4

Let $\mathscr{D} \subset \mathbf{R}^2$ be a domain; in what follows, $\mathscr{D}$ may be replaced by a two-dimensional orientable manifold (for theorems deduced by using integral formulas) or it should be bounded (in the case we are going to use the theory of pseudoanalytic functions). Let $M: \mathscr{D} \to E^4$ be a surface. To each point $m \in M$, let us associate an orthonormal frame $\{m, v_1, v_2, v_3, v_4\}$ such that $v_1, v_2 \in T_m(M)$; the vectors v_3, v_4 belong then to the *normal space* $N_m(M)$ of M at m; the totality of the spaces $N_m(M)$ is called the *normal bundle* of M. We have

$$(1.1)\qquad \begin{aligned} \mathrm{d}m &= \omega^1 v_1 + \omega^2 v_2, \\ \mathrm{d}v_1 &= \qquad\qquad \omega_1{}^2 v_2 + \omega_1{}^3 v_3 + \omega_1{}^4 v_4, \\ \mathrm{d}v_2 &= -\omega_1{}^2 v_1 \qquad\quad + \omega_2{}^3 v_3 + \omega_2{}^4 v_4, \\ \mathrm{d}v_3 &= -\omega_1{}^3 v_1 - \omega_2{}^3 v_2 \qquad\quad + \omega_3{}^4 v_4, \\ \mathrm{d}v_4 &= -\omega_1{}^4 v_1 - \omega_2{}^4 v_2 - \omega_3{}^4 v_3; \end{aligned}$$

from

$$(1.2)\qquad \omega^3 = \omega^4 = 0,$$

we get by the standard prolongation procedure the existence of functions $a_1, \ldots, a_3$, $b_1, \ldots, b_3, \alpha_1, \ldots, \alpha_4, \beta_1, \ldots, \beta_4$

$$(1.3)\qquad \begin{aligned} \omega_1{}^3 &= a_1\omega^1 + a_2\omega^2, \qquad \omega_2{}^3 = a_2\omega^1 + a_3\omega^2, \\ \omega_1{}^4 &= b_1\omega^1 + b_2\omega^2, \qquad \omega_2{}^4 = b_2\omega^1 + b_3\omega^2; \end{aligned}$$

$$(1.4)\qquad \begin{aligned} &\mathrm{d}a_1 - 2a_2\omega_1{}^2 - b_1\omega_3{}^4 = \alpha_1\omega^1 + \alpha_2\omega^2, \\ &\mathrm{d}a_2 + (a_1 - a_3)\,\omega_1{}^2 - b_2\omega_3{}^4 = \alpha_2\omega^1 + \alpha_3\omega^2, \\ &\mathrm{d}a_3 + 2a_2\omega_1{}^2 - b_3\omega_3{}^4 = \alpha_3\omega^1 + \alpha_4\omega^2, \\ &\mathrm{d}b_1 - 2b_2\omega_1{}^2 + a_1\omega_3{}^4 = \beta_1\omega^1 + \beta_2\omega^2, \\ &\mathrm{d}b_2 + (b_1 - b_3)\,\omega_1{}^2 + a_2\omega_3{}^4 = \beta_2\omega^1 + \beta_3\omega^2, \\ &\mathrm{d}b_3 + 2b_2\omega_1{}^2 + a_3\omega_3{}^4 = \beta_3\omega^1 + \beta_4\omega^2. \end{aligned}$$

Let $\{m, {}^*v_1, \ldots, {}^*v_4\}$ be another field of associated frames, and let

$$(1.5)\qquad \begin{aligned} &v_1 = \varepsilon_1 \cos\varrho \cdot {}^*v_1 - \sin\varrho \cdot {}^*v_2, \qquad v_2 = \varepsilon_1 \sin\varrho \cdot {}^*v_1 + \cos\varrho \cdot {}^*v_2, \\ &v_3 = \varepsilon_2 \cos\sigma \cdot {}^*v_3 - \sin\sigma \cdot {}^*v_4, \qquad v_4 = \varepsilon_2 \sin\sigma \cdot {}^*v_3 + \cos\sigma \cdot {}^*v_4, \\ &\varepsilon_1{}^2 = \varepsilon_2{}^2 = 1. \end{aligned}$$

We have

$$\text{(1.6)}\qquad \begin{aligned}
\mathrm{d}m &= {}^*\omega^1 \cdot {}^*v_1 + {}^*\omega^2 \cdot {}^*v_2,\\
\mathrm{d}^*v_1 &= \qquad\qquad {}^*\omega_1{}^2 \cdot {}^*v_2 + {}^*\omega_1{}^3 \cdot {}^*v_3 + {}^*\omega_1{}^4 \cdot {}^*v_4,\\
\mathrm{d}^*v_2 &= -{}^*\omega_1{}^2 \cdot {}^*v_1 \qquad + {}^*\omega_2{}^3 \cdot {}^*v_3 + {}^*\omega_2{}^4 \cdot {}^*v_4,\\
\mathrm{d}^*v_3 &= -{}^*\omega_1{}^3 \cdot {}^*v_1 - {}^*\omega_2{}^3 \cdot {}^*v_2 \qquad + {}^*\omega_3{}^4 \cdot {}^*v_4,\\
\mathrm{d}^*v_4 &= -{}^*\omega_1{}^4 \cdot {}^*v_1 - {}^*\omega_2{}^4 \cdot {}^*v_2 - {}^*\omega_3{}^4 \cdot {}^*v_3;
\end{aligned}$$

it is easy to see that

$$\text{(1.7)}\qquad {}^*\omega^1 = \varepsilon_1(\cos\varrho \cdot \omega^1 + \sin\varrho \cdot \omega^2), \quad {}^*\omega^2 = -\sin\varrho \cdot \omega^1 + \cos\varrho \cdot \omega^2;$$

$$\text{(1.8)}\qquad {}^*\omega_1{}^2 = \varepsilon_1(\mathrm{d}\varrho + \omega_1{}^2), \quad {}^*\omega_3{}^4 = \varepsilon_2(\mathrm{d}\sigma + \omega_3{}^4);$$

$$\text{(1.9)}\qquad \begin{aligned}
{}^*a_1 &= \varepsilon_2 \cos\sigma(\cos^2\varrho \cdot a_1 + \sin 2\varrho \cdot a_2 + \sin^2\varrho \cdot a_3)\\
&\quad + \varepsilon_2 \sin\sigma(\cos^2\varrho \cdot b_1 + \sin 2\varrho \cdot b_2 + \sin^2\varrho \cdot b_3),\\
{}^*a_2 &= -\frac{1}{2}\,\varepsilon_1\varepsilon_2 \cos\sigma(\sin 2\varrho \cdot a_1 - 2\cos 2\varrho \cdot a_2 - \sin 2\varrho \cdot a_3)\\
&\quad - \frac{1}{2}\,\varepsilon_1\varepsilon_2 \sin\sigma(\sin 2\varrho \cdot b_1 - 2\cos 2\varrho \cdot b_2 - \sin 2\varrho \cdot b_3),\\
{}^*a_3 &= \varepsilon_2 \cos\sigma(\sin^2\varrho \cdot a_1 - \sin 2\varrho \cdot a_2 + \cos^2\varrho \cdot a_3)\\
&\quad + \varepsilon_2 \sin\sigma(\sin^2\varrho \cdot b_1 - \sin 2\varrho \cdot b_2 + \cos^2\varrho \cdot b_3),\\
{}^*b_1 &= -\sin\sigma(\cos^2\varrho \cdot a_1 + \sin 2\varrho \cdot a_2 + \sin^2\varrho \cdot a_3)\\
&\quad + \cos\sigma(\cos^2\varrho \cdot b_1 + \sin 2\varrho \cdot b_2 + \sin^2\varrho \cdot b_3),\\
{}^*b_2 &= \frac{1}{2}\,\varepsilon_1 \sin\sigma(\sin 2\varrho \cdot a_1 - 2\cos 2\varrho \cdot a_2 - \sin 2\varrho \cdot a_3)\\
&\quad - \frac{1}{2}\,\varepsilon_1 \cos\sigma(\sin 2\varrho \cdot b_1 - 2\cos 2\varrho \cdot b_2 - \sin 2\varrho \cdot b_3),\\
{}^*b_3 &= -\sin\sigma(\sin^2\varrho \cdot a_1 - \sin 2\varrho \cdot a_2 + \cos^2\varrho \cdot a_3)\\
&\quad + \cos\sigma(\sin^2\varrho \cdot b_1 - \sin 2\varrho \cdot b_2 + \cos^2\varrho \cdot b_3).
\end{aligned}$$

From (1.4) and (1.9),

$$\text{(1.10)}\qquad \begin{aligned}
\mathrm{d}(a_1 + a_3) - (b_1 + b_3)\,\omega_3{}^4 &= (\alpha_1 + \alpha_3)\,\omega^1 + (\alpha_2 + \alpha_4)\,\omega^2,\\
\mathrm{d}(b_1 + b_3) + (a_1 + a_3)\,\omega_3{}^4 &= (\beta_1 + \beta_3)\,\omega^1 + (\beta_2 + \beta_4)\,\omega^2;
\end{aligned}$$

$$\text{(1.11)}\qquad \begin{aligned}
{}^*a_1 + {}^*a_3 &= \varepsilon_2 \cos\sigma \cdot (a_1 + a_3) + \varepsilon_2 \sin\sigma \cdot (b_1 + b_3),\\
{}^*b_1 + {}^*b_3 &= -\sin\sigma \cdot (a_1 + a_3) + \cos\sigma \cdot (b_1 + b_3);
\end{aligned}$$

from these,

$$\begin{aligned}(1.12)\quad \varepsilon_1\varepsilon_2(*\alpha_1 + *\alpha_3) &= \cos\varrho\cos\sigma\cdot(\alpha_1+\alpha_3) + \sin\varrho\cos\sigma\cdot(\alpha_2+\alpha_4)\\ &\quad + \cos\varrho\sin\sigma\cdot(\beta_1+\beta_3) + \sin\varrho\sin\sigma\cdot(\beta_2+\beta_4),\\ \varepsilon_2(*\alpha_2 + *\alpha_4) &= -\sin\varrho\cos\sigma\cdot(\alpha_1+\alpha_3) + \cos\varrho\cos\sigma\cdot(\alpha_2+\alpha_4)\\ &\quad - \sin\varrho\sin\sigma\cdot(\beta_1+\beta_3) + \cos\varrho\sin\sigma\cdot(\beta_2+\beta_4),\\ \varepsilon_1(*\beta_1 + *\beta_3) &= -\cos\varrho\sin\sigma\cdot(\alpha_1+\alpha_3) - \sin\varrho\sin\sigma\cdot(\alpha_2+\alpha_4)\\ &\quad + \cos\varrho\cos\sigma\cdot(\beta_1+\beta_3) + \sin\varrho\cos\sigma\cdot(\beta_2+\beta_4),\\ *\beta_2 + *\beta_4 &= \sin\varrho\sin\sigma\cdot(\alpha_1+\alpha_3) - \cos\varrho\sin\sigma\cdot(\alpha_2+\alpha_4)\\ &\quad - \sin\varrho\cos\sigma\cdot(\beta_1+\beta_3) + \cos\varrho\cos\sigma\cdot(\beta_2+\beta_4).\end{aligned}$$

Lemma 1.1. *Introduce the normal vector field*

$$(1.13)\quad \xi = (a_1 + a_3)\, v_3 + (b_1 + b_3)\, v_4$$

and the functions

$$(1.14)\quad H = (a_1 + a_3)^2 + (b_1 + b_3)^2,$$

$$(1.15)\quad K = a_1a_3 - a_2^2 + b_1b_3 - b_2^2,$$

$$(1.16)\quad k = (a_1 - a_3)\, b_2 - (b_1 - b_3)\, a_2 .$$

Then

$$(1.17)\quad *\xi = \xi$$

and

$$(1.18)\quad *K = K, \quad *H = H, \quad *k = \varepsilon_1\varepsilon_2 k .$$

The proof is an easy consequence of (1.5) and (1.9).

Definition 1.1. *The vector ξ is the so-called* mean curvature vector *and* (1.14) *to* (1.16) *are the* mean curvature, Gauss' curvature *and the* curvature of the normal bundle *resp.*

Notice that

$$(1.19)\quad d\omega_1^{\ 2} = -K\omega^1 \wedge \omega^2, \quad d\omega_3^{\ 4} = -k\omega^1 \wedge \omega^2 .$$

Definition 1.2. *Let n be a normal vector field on M. This field is called* parallel *if $tn \in T_m(M)$ for each vector $t \in T_m(M)$ and for each point $m \in M$.*

Lemma 1.2. *The following conditions are equivalent:* (i) $k = 0$ *on* M; (ii) *let* $m \in M$ *and let $v_0 \in N_m(M)$ be an arbitrary normal vector; then there is a neighbourhood $U \subset M$ of m and a parallel normal vector field v over U such that $v(m) = v_0$.*

Proof. Let

$$(1.20)\quad n = xv_3 + yv_4$$

be a normal vector field. Then

$$\begin{aligned}(1.21)\quad dn = &-(x\omega_1^{\ 3} + y\omega_1^{\ 4})\, v_1 - (x\omega_2^{\ 3} + y\omega_2^{\ 4})\, v_2 + (dx - y\omega_3^{\ 4})\, v_3\\ &+ (dy + x\omega_3^{\ 4})\, v_4,\end{aligned}$$

i.e.,

$$(1.22)\quad tn = -\{x\omega_1{}^3(t) + y\omega_1{}^4(t)\}\, v_1 - \{x\omega_2{}^3(t) + y\omega_2{}^4(t)\}\, v_2 + \{tx - y\omega_3{}^4(t)\}\, v_3 + \{ty + x\omega_3{}^4(t)\}\, v_4 .$$

The vector field (1.20) being parallel, we have

$$(1.23)\quad \mathrm{d}x - y\omega_3{}^4 = 0, \qquad \mathrm{d}y + x\omega_3{}^4 = 0.$$

Because of (1.19), (1.23) imply

$$(1.24)\quad yk\omega^1 \wedge \omega^2 = xk\omega^1 \wedge \omega^2 = 0$$

and $k = 0$ because of $x^2 + y^2 \neq 0$ which may be supposed at m. Thus (ii) implies (i). On the contrary, (i) implies the complete integrability of (1.23), and we get (ii). QED.

The linear mapping

$$(1.25)\quad C_m : T_m(M) \to N_m(M), \qquad m \in M,$$

be defined by

$$(1.26)\quad C_m(t) = (t\xi)^N, \qquad t \in T_m(M),$$

η^N being the normal component of the vector η and ξ the mean curvature vector. From

$$(1.27)\quad \mathrm{d}\xi = -\{(a_1 + a_3)(a_1\omega^1 + a_2\omega^2) + (b_1 + b_3)(b_1\omega^1 + b_2\omega^2)\}\, v_1 - \{(a_1 + a_3)(a_2\omega^1 + a_3\omega^2) + (b_1 + b_3)(b_2\omega^1 + b_3\omega^2)\}\, v_2 + \{(\alpha_1 + \alpha_3)\,\omega^1 + (\alpha_2 + \alpha_4)\,\omega^2\}\, v_3 + \{(\beta_1 + \beta_3)\,\omega^1 + (\beta_2 + \beta_4)\,\omega^2\} v_4 ,$$

we get

$$(1.28)\quad C(v_1) = (\alpha_1 + \alpha_3)\, v_3 + (\beta_1 + \beta_3)\, v_4, \quad C(v_2) = (\alpha_2 + \alpha_4)\, v_3 + (\beta_2 + \beta_4)\, v_4 .$$

Lemma 1.3. *Let*

$$(1.29)\quad \Gamma = \begin{vmatrix} \alpha_1 + \alpha_3 & \alpha_2 + \alpha_4 \\ \beta_1 + \beta_3 & \beta_2 + \beta_4 \end{vmatrix}.$$

Then

$$(1.30)\quad {}^*\Gamma = \varepsilon_1\varepsilon_2\Gamma .$$

The proof is trivial, see (1.12).

Theorem 1.1. *Let $M \subset E^4$ be a surface. Suppose:* (i) $H > 0$ *and* $k\Gamma \leqq 0$ *on* M; (ii) $C_m\big(T_m(M)\big) \subset \xi(m)$ *for each* $m \in \partial M$. *Then* $k = 0$ *on* M.

Proof. We have $\xi \neq 0$ on M because of $H \neq 0$. Let us restrict ourselves to the frames of M such that the mean curvature vector is a multiple of v_3, i.e., let us suppose

$$(1.31)\quad b_1 + b_3 = 0.$$

Choosing other frames restricted by the same condition $*b_1 + *b_3 = 0$, we get $\sin\sigma = 0$ from (1.11_2). Thus $\cos\sigma \neq 0$, $d\sigma = 0$, and (1.8_2) reduces to

$$(1.32) \qquad *\omega_3{}^4 = \varepsilon_2\omega_3{}^4 .$$

On M, consider the 1-form

$$(1.33) \qquad \tau = (a_1 + a_3)^2\, \omega_3{}^4 = (a_1 + a_3)\,\{(\beta_1 + \beta_3)\,\omega^1 + (\beta_2 + \beta_4)\,\omega^2\};$$

(1.32) implies

$$(1.34) \qquad *\tau = \varepsilon_2\tau .$$

From (ii) and (1.28) we get,

$$(1.35) \qquad \tau = 0 \quad \text{on} \quad \partial M .$$

The assertion of our theorem follows then easily from the integral formula

$$(1.36) \qquad \int_{\partial M} \tau = \int_M \{2\Gamma - (a_1 + a_3)^2\, k\}\, \mathrm{d}o .$$

QED.

This theorem opens one of the possibilities to characterize the spheres among the surfaces in E^4. Let $k = 0$ on M, and let n (1.20) be a non-trivial parallel normal vector field on M. The frames of M be chosen in such a way that v_3 and n are dependent, i.e., $y = 0$. From (1.23), $\mathrm{d}x = 0$ and $x\omega_3{}^4 = 0$, i.e.,

$$(1.37) \qquad x = \text{const} \neq 0, \qquad \omega_3{}^4 = 0,$$

and the equations (1.4) take a very simple form. The concrete consequences are left to the reader.

Let us turn our attention to the I-isometries; let us start with the infinitesimal theory.

Consider a surface $M \subset E^4$, take $m \in M$. The map

$$(1.38) \qquad II_m : N_m(M) \times T_m(M) \to \mathbf{R}$$

be defined by

$$(1.39) \qquad II_m(n_0, t) = -\langle tn, tn \rangle$$

for any local section $n: M \to N(M)$ around m such that $n(m) = n_0$. This is a good definition; indeed, let n (1.20) be a normal vector field around m, then

$$(1.40) \qquad II_m(n, .) = (xa_1 + yb_1)\,(\omega^1)^2 + 2(xa_2 + yb_2)\,\omega^1\omega^2 + (xa_3 + yb_3)\,(\omega^2)^2 .$$

Definition 1.2. *The mapping* (1.38) *is called the* second fundamental form *of* M.

Definition 1.3. *Let* $M \subset E^4$ *be a surface,* $V(E^4)$ *the vector space of* E^4, $v: M \to V(E^4)$ *a vector field. The field* v *is called an* infinitesimal isometry *of* M *if*

$$(1.41) \qquad \langle \mathrm{d}m, \mathrm{d}v \rangle = 0, \quad \textit{i.e.,}$$

$$\langle tm, tv \rangle = 0 \quad \textit{for each} \quad t \in T(M).$$

This last definition is in accord with the definition of infinitesimal isometries of surfaces in E^3.

Theorem 1.2. *Let $M\colon \mathscr{D} \to E^4$, $\mathscr{D} \subset \mathbf{R}^2$ a bounded domain, be a surface. Let $n\colon M \to N(M)$ be a section such that, for each $m \in M$, the form $II_m(n_m, .)$ is definite and the vector n_m is not orthogonal to the mean curvature vector ξ_m at m. Let v be an infinitesimal isometry of M such that, again for each $m \in M$, the vector v_m is situated in the vector space spanned by $T_m(M)$ and n_m. Further, let v_m be orthogonal to $T_m(M)$ for each $m \in \partial M$. Then $v = 0$ on M.*

Proof. Let us choose the frames of M as above, and let

$$v = Av_1 + Bv_2 + Cv_3 + Dv_4. \tag{1.42}$$

Then

$$\begin{aligned} dv = {} & (dA - B\omega_1{}^2 - C\omega_1{}^3 - D\omega_1{}^4)\, v_1 + (dB + A\omega_1{}^2 - C\omega_2{}^3 - D\omega_2{}^4)\, v_2 \\ & + (dC + A\omega_1{}^3 + B\omega_2{}^3 - D\omega_3{}^4)\, v_3 + (dD + A\omega_1{}^4 + B\omega_2{}^4 + C\omega_3{}^4)\, v_4. \end{aligned} \tag{1.43}$$

The condition (1.41) reduces to

$$\omega^1(dA - B\omega_1{}^2 - C\omega_1{}^3 - D\omega^4) + \omega^2\,(dB + A\omega_1{}^2 - C\omega_2{}^3 - D\omega_2{}^4) = 0, \tag{1.44}$$

this implying the existence of a function P such that

$$dA - B\omega_1{}^2 - C\omega_1{}^3 - D\omega_1{}^4 = P\omega^2, \quad dB + A\omega_1{}^2 - C\omega_2{}^3 - D\omega_2{}^4 = -P\omega^1. \tag{1.45}$$

Let us write

$$n = xv_3 + yv_4. \tag{1.46}$$

The condition $\langle \xi, n \rangle \neq 0$ reads

$$(a_1 + a_3)\, x + (b_1 + b_3)\, y \neq 0. \tag{1.47}$$

The form (1.40) being definite, we have

$$(xa_1 + yb_1)\,(xa_3 + yb_3) - (xa_2 + yb_2)^2 > 0. \tag{1.48}$$

Because v is a linear combination of the vectors v_1, v_2, n, there exists a function Q such that

$$C = xQ, \qquad D = yQ, \tag{1.49}$$

and the equations (1.45) may be rewritten as

$$\begin{aligned} dA - B\omega_1{}^2 &= (xa_1 + yb_1)\, Q\omega^1 + \{(xa_2 + yb_2)\, Q + P\}\, \omega^2, \\ dB + A\omega_1{}^2 &= \{(xa_2 + yb_2)\, Q - P\}\, \omega^1 + (xa_3 + yb_3)\, Q\omega^2. \end{aligned} \tag{1.50}$$

On M, consider, as usually, isothermic coordinates (u, v), and the frames of M be chosen in such a way that

$$\omega^1 = r\, du, \quad \omega^2 = r\, dv, \quad \omega_1{}^2 = \frac{1}{r}\left(-\frac{\partial r}{\partial v}\, du + \frac{\partial r}{\partial u}\, dv\right); \quad r = r(u, v) > 0. \tag{1.51}$$

Then (1.50) imply

$$\begin{aligned} & \frac{\partial A}{\partial u} + \frac{1}{r}\frac{\partial r}{\partial v} B = (xa_1 + yb_1)\, Qr, \qquad \frac{\partial A}{\partial v} - \frac{1}{r}\frac{\partial r}{\partial u} B = (xa_2 + yb_2)\, Qr + Pr, \\ & \frac{\partial B}{\partial u} - \frac{1}{r}\frac{\partial r}{\partial v} A = (xa_2 + yb_2)\, Qr - Pr, \quad \frac{\partial B}{\partial v} + \frac{1}{r}\frac{\partial r}{\partial u} A = (xa_3 + yb_3)\, Qr. \end{aligned} \tag{1.52}$$

The elimination of P and Q yields the system

$$(1.53)\quad (xa_3 + yb_3)\frac{\partial A}{\partial u} - (xa_1 + yb_1)\frac{\partial B}{\partial v} = (xa_1 + yb_1)\frac{1}{r}\frac{\partial r}{\partial u}A - (xa_3 + yb_3)\frac{1}{r}\frac{\partial r}{\partial v}B,$$

$$2(xa_2 + yb_2)\left(\frac{\partial A}{\partial u} + \frac{\partial B}{\partial v}\right) - \{x(a_1 + a_3) + y(b_1 + b_3)\}\left(\frac{\partial A}{\partial v} + \frac{\partial B}{\partial u}\right)$$
$$= -2(xa_2 + yb_2)\frac{1}{r}\left(\frac{\partial r}{\partial u}A + \frac{\partial r}{\partial v}B\right) - \{x(a_1 + a_3) + y(b_1 + b_3)\}\frac{1}{r}\left(\frac{\partial r}{\partial v}A + \frac{\partial r}{\partial u}B\right).$$

The discriminant of the associated form of this system is equal to

$$(1.54)\quad \Delta = 4\{(xa_1 + yb_1)(xa_3 + yb_3) - (xa_2 + yb_2)^2\} \cdot \{x(a_1 + a_3) + y(b_1 + b_3)\}^2;$$

because of (1.47) and (1.48), $\Delta > 0$, and the system (1.53) is elliptic. On ∂M, $A = B = 0$; thus $A = B = 0$ on M. The equations (1.50) reduce to

$$(1.55)\quad (xa_1 + yb_1)\,Q = (xa_2 + yb_2)\,Q = (xa_3 + yb_3)\,Q = 0;$$

thus $Q = 0$ and $C = D = 0$. QED.

This theorem is a direct generalization of Theorem VI.2.2. Indeed, let $M \subset E^3 \subset E^4$ be a surface and $v\colon M \to V(E^3)$ its infinitesimal isometry. The section n be chosen as the unit normal vector field of M. Suppose $K > 0$ on M. Then the form $II_m(n_m, .)$ reduces to the second fundamental form of M ($K > 0$ implying its definiteness), and the mean curvature vector field is a non-zero multiple of n. Thus all conditions of Theorem 1.2 are fulfilled, and $v = 0$ on M follows.

Let us turn our attention to isometric surfaces in E^4. Two isometric surfaces are globally non-equivalent even under suitable conditions on the positiveness of the curvatures. To ensure the equivalence, we have to add a further condition, this condition being automatically satisfied for surfaces in E^3.

Let $M\colon \mathscr{D} \to E^4$ be a surface and let $M^*\colon \mathscr{D} \to E^4$ be another surface isometric to M. Then we are in the position to choose the frames $\{m^*, v_1^*, v_2^*, v_3^*, v_4^*\}$ of M^* in such a way that

$$(1.56)\quad \begin{aligned} dm^* &= \omega^1 v_1^* + \omega^2 v_2^*,\\ dv_1^* &= (\omega_1^2 + \tau_1^2)\,v_2^* + (\omega_1^3 + \tau_1^3)\,v_3^* + (\omega_1^4 + \tau_1^4)\,v_4^*,\\ dv_2^* &= -(\omega_1^2 + \tau_1^2)\,v_1^* + (\omega_2^3 + \tau_2^3)\,v_3^* + (\omega_2^4 + \tau_2^4)\,v_4^*,\\ dv_3^* &= -(\omega_1^3 + \tau_1^3)\,v_1^* - (\omega_2^3 + \tau_2^3)\,v_2^* + (\omega_3^4 + \tau_3^4)\,v_4^*,\\ dv_4^* &= -(\omega_1^4 + \tau_1^4)\,v_1^* - (\omega_2^4 + \tau_2^4)\,v_2^* - (\omega_3^4 + \tau_3^4)\,v_3^*. \end{aligned}$$

From the integrability conditions of (1.56), we get

$$(1.57)\quad \omega^1 \wedge \tau_1^2 = 0, \quad \omega^2 \wedge \tau_1^2 = 0,$$

$$(1.58)\quad \omega^1 \wedge \tau_1^3 + \omega^2 \wedge \tau_2^3 = 0, \quad \omega^1 \wedge \tau_1^4 + \omega^2 \wedge \tau_2^4 = 0,$$

$$(1.59)\quad d\tau_1^2 = -\omega_1^3 \wedge \tau_2^3 - \tau_1^3 \wedge \omega_2^3 - \tau_1^3 \wedge \tau_2^3 - \omega_1^4 \wedge \tau_2^4 - \tau_1^4 \wedge \omega_2^4 - \tau_1^4 \wedge \tau_2^4,$$

$$(1.60)\quad \begin{aligned} d\tau_1^3 &= \omega_1^2 \wedge \tau_2^3 + \tau_1^2 \wedge \omega_2^3 + \tau_1^2 \wedge \tau_2^3 - \omega_1^4 \wedge \tau_3^4 - \tau_1^4 \wedge \omega_3^4 - \tau_1^4 \wedge \tau_3^4, \\ d\tau_2^3 &= -\omega_1^2 \wedge \tau_1^3 - \tau_1^2 \wedge \omega_1^3 - \tau_1^2 \wedge \tau_1^3 - \omega_2^4 \wedge \tau_3^4 - \tau_2^4 \wedge \omega_3^4 - \tau_2^4 \wedge \tau_3^4, \\ d\tau_1^4 &= \omega_1^2 \wedge \tau_2^4 + \tau_1^2 \wedge \omega_2^4 + \tau_1^2 \wedge \tau_2^4 + \omega_1^3 \wedge \tau_3^4 + \tau_1^3 \wedge \omega_3^4 + \tau_1^3 \wedge \tau_3^4, \\ d\tau_2^4 &= -\omega_1^2 \wedge \tau_1^4 - \tau_1^2 \wedge \omega_1^4 - \tau_1^2 \wedge \tau_1^4 + \omega_2^3 \wedge \tau_3^4 + \tau_2^3 \wedge \omega_3^4 + \tau_2^3 \wedge \tau_3^4, \\ d\tau_3^4 &= -\omega_1^3 \wedge \tau_1^4 - \tau_1^3 \wedge \omega_1^4 - \tau_1^3 \wedge \tau_1^4 - \omega_2^3 \wedge \tau_2^4 - \tau_2^3 \wedge \omega_2^4 - \tau_2^3 \wedge \tau_2^4. \end{aligned}$$

From (1.57),

$$(1.61)\quad \tau_1^2 = 0,$$

and the equation (1.59) reduces to

$$(1.62)\quad \omega_1^3 \wedge \tau_2^3 + \tau_1^3 \wedge \omega_2^3 + \tau_1^3 \wedge \tau_2^3 + \omega_1^4 \wedge \tau_2^4 + \tau_1^4 \wedge \omega_2^4 + \tau_1^4 \wedge \tau_2^4 = 0.$$

From (1.58), we get the existence of functions $R_1, \ldots, S_3 : \mathscr{D} \to \mathbf{R}$ such that

$$(1.63)\quad \begin{aligned} \tau_1^3 &= R_1\omega^1 + R_2\omega^2, & \tau_1^4 &= S_1\omega^1 + S_2\omega^2, \\ \tau_2^3 &= R_2\omega^1 + R_3\omega^2, & \tau_2^4 &= S_2\omega^1 + S_3\omega^2; \end{aligned}$$

by the standard prolongation, we get the existence of functions $R_1^*, \ldots, S_4^* : \mathscr{D} \to \mathbf{R}$ such that

$$(1.64)\quad \begin{aligned} &dR_1 - 2R_2\omega_1^2 - b_1\tau_3^4 - S_1(\omega_3^4 + \tau_3^4) = R_1^*\omega^1 + R_2^*\omega^2, \\ &dR_2 + (R_1 - R_3)\,\omega_1^2 - b_2\tau_3^4 - S_2(\omega_3^4 + \tau_3^4) = R_2^*\omega^1 + R_3^*\omega^2, \\ &dR_3 + 2R_2\omega_1^2 - b_3\tau_3^4 - S_3(\omega_3^4 + \tau_3^4) = R_3^*\omega^1 + R_4^*\omega^2, \\ &dS_1 - 2S_2\omega_1^2 + a_1\tau_3^4 + R_1(\omega_3^4 + \tau_3^4) = S_1^*\omega^1 + S_2^*\omega^2, \\ &dS_2 + (S_1 - S_3)\,\omega_1^2 + a_2\tau_3^4 + R_2(\omega_3^4 + \tau_3^4) = S_2^*\omega^1 + S_3^*\omega^2, \\ &dS_3 + 2S_2\omega_1^2 + a_3\tau_3^4 + R_3(\omega_3^4 + \tau_3^4) = S_3^*\omega^1 + S_4^*\omega^2. \end{aligned}$$

The equation (1.62) takes the form

$$(1.65)\quad a_3R_1 - 2a_2R_2 + a_1R_3 + R_1R_3 - R_2^2 + b_3S_1 - 2b_2S_2 + b_1S_3 + S_1S_3 - S_2^2 = 0.$$

Let $d \in \mathscr{D}$, $m_0 = M(d)$, $m_0^* = M^*(d)$. Further, let $I(d) : E^4 \to E^4$ be an isometry satisfying

$$(1.66)\quad I(d) \circ (dM^*)_d = (dM)_d.$$

Each such so-called *tangent isometry* is given by

$$(1.67)\quad \begin{aligned} &I(d)\, m_0^* = m_0, \quad I(d)\, v_1^* = v_1, \quad I(d)\, v_2^* = v_2, \\ &I(d)\, v_3^* = \varepsilon(\cos\alpha \cdot v_3 - \sin\alpha \cdot v_4), \quad I(d)\, v_4^* = \sin\alpha \cdot v_3 + \cos\alpha \cdot v_4; \\ &\varepsilon = \pm 1. \end{aligned}$$

Let $m = m(s)$ be a curve on M; suppose that s is its arc and $m_0 = m(s_0)$. Let $m^* = m^*(s)$ be the corresponding curve on M^*. Then

$$(1.68)\quad I(d)\,\frac{dm^*(s_0)}{ds} = \frac{dm(s_0)}{ds}.$$

Further,

$$(1.69)\quad I(d)\,\frac{\mathrm{d}^2m^*(s_0)}{\mathrm{d}s^2} = \frac{\mathrm{d}^2m(s_0)}{\mathrm{d}s^2} + \frac{\mathscr{L}_{I(d)}}{(\omega^1)^2+(\omega^2)^2}\left(\frac{\mathrm{d}m(s_0)}{\mathrm{d}s}\right),$$

where

$$(1.70)\quad \begin{aligned}\mathscr{L}_{I(d)} = {} & \{(\omega^1\omega_1{}^3+\omega^2\omega_2{}^3)(\varepsilon\cos\alpha-1)+(\omega^1\omega_1{}^4+\omega^2\omega_2{}^4)\sin\alpha \\ & +(\omega^1\tau_1{}^3+\omega^2\tau_2{}^3)\,\varepsilon\cos\alpha+(\omega^1\tau_1{}^4+\omega^2\tau_2{}^4)\sin\alpha\}\,v_3 \\ & +\{-(\omega^1\omega_1{}^3+\omega^2\omega_2{}^3)\,\varepsilon\sin\alpha+(\omega^1\omega_1{}^4+\omega^2\omega_2{}^4)(\cos\alpha-1) \\ & -(\omega^1\tau_1{}^3+\omega^2\tau_2{}^3)\,\varepsilon\sin\alpha+(\omega^1\tau_1{}^4+\omega^2\tau_2{}^4)\cos\alpha\}\,v_4\,.\end{aligned}$$

Thus, for each tangent isometry $I(d)$, we get the quadratic mapping

$$(1.71)\quad \mathscr{L}_{I(d)}: T_{m_0}(M) \to N_{m_0}(M),$$

its geometrical interpretation being given by (1.69).

Definition 1.4. *Let $M, M^*: \mathscr{D} \to E^4$ be two isometric surfaces, $d \in \mathscr{D}$ a point and $I(d)$ a tangent isometry. Then the mapping $\mathscr{L}_{I(d)}$* (1.71) *is called the* linearity transformation.

In the connection with Definition 1.2, let us introduce the following

Definition 1.5. *Let $n \in N_m(M)$ be a normal vector at $m \in M$, let $II(n, .)$* (1.40) *be the second fundamental form of M. Then the invariants*

$$(1.72)\quad K_n = (xa_1+yb_1)(xa_3+yk_3)-(xa_2+yb_2)^2,$$

$$(1.73)\quad 2H_n = x(a_1+a_3)+y(b_1+b_3)$$

are called the Gauss *and* mean curvatures of M with respect to n.

Theorem 1.3. *Let $M, M^*: \mathscr{D} \to E^4$, $\mathscr{D} \subset \mathbf{R}^2$ a bounded domain, be two isometric surfaces. Let there exist a section $n: M \to N(M)$, n a unit vector, such that, for each $d \in \mathscr{D} \cup \partial\mathscr{D}$, there is a tangent isometry $I(d)$ satisfying*

$$(1.74)\quad \mathscr{L}_{I(d)}\big(T_{M(d)}(M)\big) \subset n(d) \quad \text{on} \quad \mathscr{D} \cup \partial\mathscr{D}$$

and

$$(1.75)\quad \mathscr{L}_{I(d)}\big(T_{M(d)}(M)\big) = 0 \quad \text{on} \quad \partial\mathscr{D}.$$

Further, suppose

$$(1.76)\quad K_n > 0,\quad H_n \neq 0;\quad H_n + H^*_{I(d)^{-1}n} \neq 0$$

on $\mathscr{D}$. Then M and M^* are equivalent (i.e., they are the same up to a motion of E^4).

Proof. The frames be chosen in such a way that

$$(1.77)\quad n = v_3,\quad I(d)\,v_3{}^* = v_3,\quad I(d)\,v_4{}^* = v_4.$$

Then

$$(1.78)\quad \begin{aligned}\mathscr{L}_{I(d)} = {} & \{R_1(\omega^1)^2+2R_2\omega^1\omega^2+R_3(\omega^2)^2\}\,v_3 \\ & +\{S_1(\omega^1)^2+2S_2\omega^1\omega^2+S_3(\omega^2)^2\}\,v_4\,.\end{aligned}$$

Our suppositions read:

$$S_1 = S_2 = S_3 = 0 \quad \text{on } \mathscr{D} \cup \partial\mathscr{D}, \tag{1.79}$$

$$R_1 = R_2 = R_3 = 0 \quad \text{on } \partial\mathscr{D}, \tag{1.80}$$

$$a_1a_3 - a_2^2 > 0, \quad a_1 + a_3 \neq 0, \quad 2(a_1 + a_3) + R_1 + R_3 \neq 0 \quad \text{on } \mathscr{D}. \tag{1.81}$$

The relation (1.65) may be rewritten as

$$(2a_3 + R_3) R_1 - 2(2a_2 + R_2) R_2 + (2a_1 + R_1) R_3 = 0, \tag{1.82}$$

and we get

$$\begin{aligned} &R_1 = A_1(R_1 - R_3) + A_2R_2, \quad R_3 = A_3(R_1 - R_3) + A_2R_2; \\ &A_1 := A(2a_1 + R_1), \quad A_2 := 2A(2a_2 + R_2), \quad A_3 := -A(2a_3 + R_3), \\ &A := (2a_1 + 2a_3 + R_1 + R_3)^{-1}. \end{aligned} \tag{1.83}$$

Let us write

$$\omega_3^4 = c_1\omega^1 + c_2\omega^2, \qquad \tau_3^4 = x\omega^1 + y\omega^2. \tag{1.84}$$

From (1.64_{4-6}), we get

$$\begin{aligned} &(a_2 + R_2)\, x - (a_1 + R_1)\, y = c_2R_1 - c_1R_2, \\ &(a_3 + R_3)\, x - (a_2 + R_2)\, y = c_2R_2 - c_1R_3, \end{aligned} \tag{1.85}$$

i.e.,

$$\begin{aligned} &x = B_1(R_1 - R_3) + B_2R_2, \quad y = B_3(R_1 - R_3) + B_4R_2; \\ &B_1 := -B\{c_2(a_2 + R_2)\, A_1 + c_1(a_1 + R_1)\, A_3\}, \\ &B_2 := B\{(a_2 + R_2)\,(c_1 - c_2A_2) + (a_1 + R_1)\,(c_2 - c_1A_2)\}, \\ &B_3 := -B\{c_2(a_3 + R_3)\, A_1 + c_1(a_2 + R_2)\, A_3\}, \\ &B_4 := B\{(a_2 + R_2)\,(c_2 - c_1A_2) + (a_3 + R_3)\,(c_1 - c_2A_2)\}, \\ &B := (a_1a_3 - a_2^2)^{-1}. \end{aligned} \tag{1.86}$$

From (1.64), we have

$$\begin{aligned} &\mathrm{d}(R_1 - R_3) - 4R_2\omega_1^2 - (b_1 - b_3)\, \tau_3^4 = (R_1^* - R_3^*)\, \omega^1 + (R_2^* - R_4^*)\,\omega^2, \\ &\mathrm{d}R_2 + (R_1 - R_3)\, \omega_1^2 - b_2\tau_3^4 = R_2^*\omega^1 + R_3^*\omega^2. \end{aligned} \tag{1.87}$$

On $\mathscr{D}$, introduce isothermic coordinates (u, v), and let us choose the vector fields v_1, v_2 in such a way that

$$\begin{aligned} &\omega^1 = r\,\mathrm{d}u, \quad \omega^2 = r\,\mathrm{d}v, \quad \omega_1^2 = \frac{1}{r}\left(-\frac{\partial r}{\partial v}\,\mathrm{d}u + \frac{\partial r}{\partial u}\,\mathrm{d}v\right); \\ &r = r(u, v) > 0. \end{aligned} \tag{1.88}$$

The equations (1.87) imply

$$(1.89)\quad \frac{\partial(R_1 - R_3)}{\partial u} = (R_1{}^* - R_3{}^*)\, r + C_1(R_1 - R_3) + C_2 R_2,$$

$$\frac{\partial(R_1 - R_3)}{\partial v} = (R_2{}^* - R_4{}^*)\, r + C_3(R_1 - R_3) + C_4 R_2,$$

$$\frac{\partial R_2}{\partial u} = R_2{}^* r + C_5(R_1 - R_3) + C_6 R_2,$$

$$\frac{\partial R_2}{\partial v} = R_3{}^* r + C_7(R_1 - R_3) + C_8 R_2;$$

$$C_1 := B_1(b_1 - b_3)\, r, \quad C_2 := B_2(b_1 - b_3)\, r - 4\frac{1}{r}\frac{\partial r}{\partial v},$$

$$C_3 := B_3(b_1 - b_3)\, r, \quad C_4 := B_4(b_1 - b_3)\, r + 4\frac{1}{r}\frac{\partial r}{\partial u},$$

$$C_5 := B_1 b_2 r + \frac{1}{r}\frac{\partial r}{\partial v}, \quad C_6 := B_2 b_2 r,$$

$$C_7 := B_3 b_2 r - \frac{1}{r}\frac{\partial r}{\partial u}, \quad C_8 := B_4 b_2 r.$$

From (1.82), we get

$$(1.90)\quad a_3 R_1{}^* - 2a_2 R_2{}^* + a_1 R_3{}^*$$
$$= -(b_3 c_1 + \alpha_3 + b_3 x + R_3{}^*)\, R_1 + 2(b_2 c_1 + \alpha_2 + b_2 x + R_2{}^*)\, R_2$$
$$- (b_1 c_1 + \alpha_1 + b_1 x + R_1{}^*)\, R_3 - (a_3 b_1 - 2a_2 b_2 + a_1 b_3)\, x,$$

$$a_3 R_2{}^* - 2a_2 R_3{}^* + a_1 R_4{}^*$$
$$= -(b_3 c_2 + \alpha_4 + b_3 y + R_4{}^*)\, R_1 + 2(b_2 c_2 + \alpha_3 + b_2 y + R_3{}^*)\, R_2$$
$$- (b_1 c_2 + \alpha_2 + b_1 y + R_2{}^*)\, R_3 - (a_3 b_1 - 2a_2 b_2 + a_1 b_3)\, y.$$

According to (1.83) and (1.86), these may be rewritten as

$$(1.91)\quad a_3 R_1{}^* - 2a_2 R_2{}^* + a_1 R_3{}^* = D_1(R_1 - R_3) + D_2 R_2,$$
$$a_3 R_2{}^* - 2a_2 R_3{}^* + a_1 R_4{}^* = D_3(R_1 - R_3) + D_4 R_2.$$

From (1.89) and (1.91), we get

$$(1.92)\quad a_3 \frac{\partial(R_1 - R_3)}{\partial u} - 2a_2 \frac{\partial R_2}{\partial u} + (a_1 + a_3) \frac{\partial R_2}{\partial v}$$
$$= \{rD_1 + a_3(C_1 + C_7) - 2a_2 C_5 + a_1 C_7\}\, (R_1 - R_3)$$
$$+ \{rD_2 + a_3(C_2 + C_8) - 2a_2 C_6 + a_1 C_8\}\, R_2,$$

$$-a_1 \frac{\partial(R_1 - R_3)}{\partial v} + (a_1 + a_3) \frac{\partial R_2}{\partial u} - 2a_2 \frac{\partial R_2}{\partial v}$$
$$= \{rD_3 + a_3 C_5 - 2a_2 C_7 + a_1(C_5 - C_3)\}\, (R_1 - R_3)$$
$$+ \{rD_4 + a_3 C_6 - 2a_2 C_8 + a_1(C_6 - C_4)\}\, R_2.$$

The associated form of the system (1.92) being equal to

(1.93) $\Phi = (a_1 + a_3)(a_1\mu^2 + 2a_2\mu\nu + a_3\nu^2)$,

the system (1.92) is elliptic. Thus $R_1 - R_3 = R_2 = 0$ in $\mathscr{D}$. From (1.79), $\tau_1{}^3 = \tau_2{}^3 = 0$, (1.83) implies $\tau_1{}^4 = \tau_2{}^4 = 0$, and, finally, we get $\tau_3{}^4 = 0$ from (1.84$_2$) and (1.86). QED.

2. Surfaces in E^5

Let $M: \mathscr{D} \to E^5$ be a surface. To each point $m \in M$, let us associate an orthonormal frame $\{m, v_1, \ldots, v_5\}$ such that $v_1, v_2 \in T_m(M)$. Then

(2.1)
$$\begin{aligned}
dm &= \omega^1 v_1 + \omega^2 v_2, \\
dv_1 &= \omega_1{}^2 v_2 + \omega_1{}^3 v_3 + \omega_1{}^4 v_4 + \omega_1{}^5 v_5, \\
dv_2 &= -\omega_1{}^2 v_1 + \omega_2{}^3 v_3 + \omega_2{}^4 v_4 + \omega_2{}^5 v_5, \\
dv_3 &= -\omega_1{}^3 v_1 - \omega_2{}^3 v_2 + \omega_3{}^4 v_4 + \omega_3{}^5 v_5, \\
dv_4 &= -\omega_1{}^4 v_1 - \omega_2{}^4 v_2 - \omega_3{}^4 v_3 + \omega_4{}^5 v_5, \\
dv_5 &= -\omega_1{}^5 v_1 - \omega_2{}^5 v_2 - \omega_3{}^5 v_3 - \omega_4{}^5 v_4.
\end{aligned}$$

From

(2.2) $\omega^3 = \omega^4 = \omega^5 = 0$,

we get

(2.3)
$$\omega^1 \wedge \omega_1{}^3 + \omega^2 \wedge \omega_2{}^3 = 0, \quad \omega^1 \wedge \omega_1{}^4 + \omega^2 \wedge \omega_2{}^4 = 0,$$
$$\omega^1 \wedge \omega_1{}^5 + \omega^2 \wedge \omega_2{}^5 = 0$$

and the existence of functions $a_1, \ldots, c_3$ such that

(2.4)
$$\begin{aligned}
\omega_1{}^3 &= a_1\omega^1 + a_2\omega^2, & \omega_2{}^3 &= a_2\omega^1 + a_3\omega^2, \\
\omega_1{}^4 &= b_1\omega^1 + b_2\omega^2, & \omega_2{}^4 &= b_2\omega^1 + b_3\omega^2, \\
\omega_1{}^5 &= c_1\omega^1 + c_2\omega^2, & \omega_2{}^5 &= c_2\omega^1 + c_3\omega^2.
\end{aligned}$$

From (2.1), we obtain

(2.5) $v_1 m = v_1, \quad v_2 m = v_2$,

(2.6)
$$\begin{aligned}
v_1 v_1 m &= \omega_1{}^2(v_1)\, v_2 + a_1 v_3 + b_1 v_4 + c_1 v_5, \\
v_2 v_1 m &= \omega_1{}^2(v_2)\, v_2 + a_2 v_3 + b_2 v_4 + c_2 v_5, \\
v_1 v_2 m &= -\omega_1{}^2(v_1)\, v_1 + a_2 v_3 + b_2 v_4 + c_2 v_5, \\
v_2 v_2 m &= -\omega_1{}^2(v_2)\, v_1 + a_3 v_3 + b_3 v_4 + c_3 v_5.
\end{aligned}$$

Thus the *osculating space* $T_m{}^2(M)$ of M at $m \in M$ is the space going through m and spanned by the vectors

(2.7) $v_1, \quad v_2, \quad a_1 v_3 + b_1 v_4 + c_1 v_5, \quad a_2 v_3 + b_2 v_4 + c_2 v_5, \quad a_3 v_3 + b_3 v_4 + c_3 v_5$.

Let $M^*: \mathscr{D} \to E^5$ be a surface isometric to M. The frames $\{m^*, v_1^*, \ldots, v_5^*\}$ of M^* may be chosen in such a way that

$$(2.8)\quad \begin{aligned} dm^* &= \omega^1 v_1^* + \omega^2 v_2^*,\\ dv_1^* &= (\omega_1^2 + \tau_1^2)\, v_2^* + (\omega_1^3 + \tau_1^3)\, v_3^* + (\omega_1^4 + \tau_1^4)\, v_4^* + (\omega_1^5 + \tau_1^5)\, v_5^*,\\ dv_2^* &= -(\omega_1^2 + \tau_1^2)\, v_1^* + (\omega_2^3 + \tau_2^3)\, v_3^* + (\omega_2^4 + \tau_2^4)\, v_4^* + (\omega_2^5 + \tau_2^5)\, v_5^*,\\ dv_3^* &= -(\omega_1^3 + \tau_1^3)\, v_1^* - (\omega_2^3 + \tau_2^3)\, v_2^* + (\omega_3^4 + \tau_3^4)\, v_4^* + (\omega_3^5 + \tau_3^5)\, v_5^*,\\ dv_4^* &= -(\omega_1^4 + \tau_1^4)\, v_1^* - (\omega_2^4 + \tau_2^4)\, v_2^* - (\omega_3^4 + \tau_3^4)\, v_4^* + (\omega_4^5 + \tau_4^5)\, v_5^*,\\ dv_5^* &= -(\omega_1^5 + \tau_1^5)\, v_1^* - (\omega_2^5 + \tau_2^5)\, v_2^* - (\omega_3^5 + \tau_3^5)\, v_3^* - (\omega_4^5 + \tau_4^5)\, v_4^*. \end{aligned}$$

Similarly to the case of surfaces in E^4, introduce the notion of tangent isometries $I(d): E^5 \to E^5$, $d \in \mathscr{D}$. Suppose the frames v_i^* to be chosen in such a way that

$$(2.9)\quad I(d)\, v_i^* = v_i; \qquad i = 1, \ldots, 5.$$

Then

$$(2.10)\quad I(d)\, \frac{dm^*(s_0)}{ds} = \frac{dm(s_0)}{ds},$$

$$(2.11)\quad I(d)\, \frac{d^2m^*(s_0)}{ds^2} = \frac{d^2m(s_0)}{ds^2} + \frac{\mathscr{L}_{I(d)}}{(\omega^1)^2 + (\omega^2)^2} \left(\frac{dm(s_0)}{ds}\right)$$

with

$$(2.12)\quad \mathscr{L}_{I(d)} = (\omega^1\tau_1^3 + \omega^2\tau_2^3)\, v_3 + (\omega^1\tau_1^4 + \omega^2\tau_2^4)\, v_4 + (\omega^1\tau_1^5 + \omega^2\tau_2^5)\, v_5;$$

compare with (1.68)—(1.70). Of course, we make use of the relation

$$(2.13)\quad \tau_1^2 = 0.$$

Definition 2.1. *Let* $M, M^*: \mathscr{D} \to \mathbf{R}$ *be two isometric surfaces. They are said to be in a* second order deformation *if there is, for each* $d \in \mathscr{D}$, *a tangent isometry* $I(d): E^5 \to E^5$ *such that the corresponding* linearity transformation $\mathscr{L}_{I(d)}: T_{M(d)}(M) \to N_{M(d)}(M)$ *is trivial, i.e.,* $\mathscr{L}_{I(d)}(T_{M(d)}(M)) = 0$.

Let $M, M^*: \mathscr{D} \to E^5$ be in a second order deformation, let the frames v_i, v_i^* be chosen in such a way that (2.9) is the linearity transformation satisfying $\mathscr{L}_{I(d)}(T_{M(d)}(M)) = 0$. Then

$$(2.14)\quad \tau_1^3 = \tau_2^3 = \tau_1^4 = \tau_2^4 = \tau_1^5 = \tau_2^5 = 0.$$

Without entering into the details, let us consider an *infinitesimal second order deformation* of the surface $M: \mathscr{D} \to E^5$ given by

$$(2.15)\quad \begin{aligned} dm &= \omega^1 v_1 + \omega^2 v_2,\\ dv_1 &= \omega_1^2 v_2 + \omega_1^3 v_3 + \omega_1^4 v_4 + \omega_1^5 v^5,\\ dv_2 &= -\omega_1^2 v_1 + \omega_2^3 v_3 + \omega_2^4 v_4 + \omega_2^5 v_5,\\ dv_3 &= -\omega_1^3 v_1 - \omega_2^3 v_2 + (\omega_3^4 + t\varphi_3^4 + \cdots)\, v_4 + (\omega_3^5 + t\varphi_3^5 + \cdots)\, v_5,\\ dv_4 &= -\omega_1^4 v_1 - \omega_2^4 v_2 - (\omega_3^4 + t\varphi_3^4 + \cdots)\, v_3 + (\omega_4^5 + t\varphi_4^5 + \cdots)\, v_5,\\ dv_5 &= -\omega_1^5 v_1 - \omega_2^5 v_2 - (\omega_3^5 + t\varphi_3^5 + \cdots)\, v_3 - (\omega_4^5 + t\varphi_4^5 + \cdots)\, v_4. \end{aligned}$$

Comparing the terms at t in the integrability conditions of (2.15), we get

$$(2.16)\quad \begin{aligned} &\omega_1{}^4 \wedge \varphi_3{}^4 + \omega_1{}^5 \wedge \varphi_3{}^5 = 0, \quad \omega_1{}^3 \wedge \varphi_3{}^4 - \omega_1{}^5 \wedge \varphi_4{}^5 = 0, \\ &\omega_2{}^4 \wedge \varphi_3{}^4 + \omega_2{}^5 \wedge \varphi_3{}^5 = 0, \quad \omega_2{}^3 \wedge \varphi_3{}^4 - \omega_2{}^5 \wedge \varphi_4{}^5 = 0, \\ &\omega_1{}^3 \wedge \varphi_3{}^5 + \omega_1{}^4 \wedge \varphi_4{}^5 = 0, \quad \omega_2{}^3 \wedge \varphi_3{}^5 + \omega_2{}^4 \wedge \varphi_4{}^5 = 0, \end{aligned}$$

$$(2.17)\quad \begin{aligned} &\mathrm{d}\varphi_3{}^4 = -\omega_3{}^5 \wedge \varphi_4{}^5 - \varphi_3{}^5 \wedge \omega_4{}^5, \\ &\mathrm{d}\varphi_3{}^5 = \omega_3{}^4 \wedge \varphi_4{}^5 + \varphi_3{}^4 \wedge \omega_4{}^5, \quad \mathrm{d}\varphi_4{}^5 = -\omega_3{}^4 \wedge \varphi_3{}^5 - \varphi_3{}^4 \wedge \omega_3{}^5. \end{aligned}$$

Theorem 2.1. *Let $M: \mathscr{D} \to E^5$ be a surface such that* $\dim T_m{}^2(M) = 5$ *for each $m \in M$. Then each infinitesimal second order deformation of M is trivial.*

Proof. From (2.4) and (2.16), we get the existence of functions $A_1, \ldots, C_4$ such that

$$(2.18)\quad \begin{aligned} &b_1\varphi_3{}^4 + c_1\varphi_3{}^5 = A_1\omega^1 + A_2\omega^2, \quad a_1\varphi_3{}^4 - c_1\varphi_4{}^5 = B_1\omega^1 + B_2\omega^2, \\ &b_2\varphi_3{}^4 + c_2\varphi_3{}^5 = A_2\omega^1 + A_3\omega^2, \quad a_2\varphi_3{}^4 - c_2\varphi_4{}^5 = B_2\omega^1 + B_3\omega^2, \\ &b_3\varphi_3{}^4 + c_3\varphi_3{}^5 = A_3\omega^1 + A_4\omega^2, \quad a_3\varphi_3{}^4 - c_3\varphi_4{}^5 = B_3\omega^1 + B_4\omega^2, \\ &a_1\varphi_3{}^5 + b_1\varphi_4{}^5 = C_1\omega^1 + C_2\omega^2, \\ &a_2\varphi_3{}^5 + b_2\varphi_4{}^5 = C_2\omega^1 + C_3\omega^2, \\ &a_3\varphi_3{}^5 + b_3\varphi_4{}^5 = C_3\omega^1 + C_4\omega^2. \end{aligned}$$

Write

$$(2.19)\quad \varphi_3{}^4 = x_1\omega^1 + x_2\omega^2, \quad \varphi_3{}^5 = y_1\omega^1 + y_2\omega^2, \quad \varphi_4{}^5 = z_1\omega^1 + z_2\omega^2.$$

(2.18) implies

$$(2.20)\quad \begin{aligned} &b_2x_1 - b_1x_2 + c_2y_1 - c_1y_2 = 0, \quad b_3x_1 - b_2x_2 + c_3y_1 - c_2y_2 = 0, \\ &a_2x_1 - a_1x_2 - c_2z_1 + c_1z_2 = 0, \quad a_3x_1 - a_2x_2 - c_3z_1 + c_2z_2 = 0, \\ &a_2y_1 - a_1y_2 + b_2z_1 - b_1z_2 = 0, \quad a_3y_1 - a_2y_2 + b_3z_1 - b_2z_2 = 0. \end{aligned}$$

The vectors v_1v_1m, v_1v_2m, v_2v_2m — see (2.6) — being linearly independent, we may choose the frames in such a way that

$$(2.21)\quad b_1 = c_1 = c_2 = 0, \quad a_1b_2c_3 \neq 0.$$

From (2.20), we get

$$(2.22)\quad x_1 = x_2 = y_1 = y_2 = z_1 = z_2 = 0, \quad \text{i.e.,} \quad \varphi_3{}^4 = \varphi_3{}^5 = \varphi_4{}^5 = 0.$$

QED.

Let us turn our attention to surfaces $M: \mathscr{D} \to E^5$ with $\dim T_m{}^2(M) = 4$ for each $m \in M$. The frames be chosen in such a way that $v_3, v_4 \in T_m{}^2(M)$, i.e., let us always suppose

$$(2.23)\quad \omega_1{}^5 = \omega_2{}^5 = 0.$$

Let $\{m, {}^*v_i\}$ be another field of moving frames, and let

$$(2.24)\quad \begin{aligned} &v_1 = \varepsilon_1(\cos\alpha \cdot {}^*v_1 - \sin\alpha \cdot {}^*v_2), \quad v_2 = \sin\alpha \cdot {}^*v_1 + \cos\alpha \cdot v^*{}_2, \\ &v_3 = \varepsilon_2(\cos\beta \cdot {}^*v_3 - \sin\beta \cdot {}^*v_4), \quad v_4 = \sin\beta \cdot {}^*v_3 + \cos\beta \cdot {}^*v_4, \\ &v_5 = \varepsilon_3 \cdot {}^*v_5; \qquad \varepsilon_1^2 = \varepsilon_2^2 = \varepsilon_3^2 = 1. \end{aligned}$$

We get

$$(2.25)\quad \omega^1 = \varepsilon_1(\cos\alpha \cdot {}^*\omega^1 - \sin\alpha \cdot {}^*\omega^2), \quad \omega^2 = \sin\alpha \cdot {}^*\omega^1 + \cos\alpha \cdot {}^*\omega^2,$$

$$(2.26)\quad \begin{aligned} \omega_1{}^3 &= \varepsilon_1\varepsilon_2(\cos\alpha\cos\beta \cdot {}^*\omega_1{}^3 - \sin\alpha\cos\beta \cdot {}^*\omega_2{}^3 - \cos\alpha\sin\beta \cdot {}^*\omega_1{}^4 \\ &\qquad + \sin\alpha\sin\beta \cdot {}^*\omega_2{}^4), \\ \omega_1{}^4 &= \varepsilon_1(\cos\alpha\sin\beta \cdot {}^*\omega_1{}^3 - \sin\alpha\sin\beta \cdot {}^*\omega_2{}^3 + \cos\alpha\cos\beta \cdot {}^*\omega_1{}^4 \\ &\qquad - \sin\alpha\cos\beta \cdot {}^*\omega_2{}^4), \\ \omega_2{}^3 &= \varepsilon_2(\sin\alpha\cos\beta \cdot {}^*\omega_1{}^3 + \cos\alpha\cos\beta \cdot {}^*\omega_2{}^3 - \sin\alpha\sin\beta \cdot {}^*\omega_1{}^4 \\ &\qquad - \cos\alpha\sin\beta \cdot {}^*\omega_2{}^4), \\ \omega_2{}^4 &= \sin\alpha\sin\beta \cdot {}^*\omega_1{}^3 + \cos\alpha\sin\beta \cdot {}^*\omega_2{}^3 + \sin\alpha\cos\beta \cdot {}^*\omega_1{}^4 \\ &\qquad + \cos\alpha\cos\beta \cdot {}^*\omega_2{}^4. \end{aligned}$$

By an easy calculation, we get

Lemma 2.1. *Consider the functions*

$$(2.27)\quad K = a_1a_3 - a_2^2 + b_1b_3 - b_2^2, \qquad k = (a_1 - a_3)\, b_2 - a_2(b_1 - b_3).$$

Then

$$(2.28)\quad {}^*K = K, \qquad {}^*k = \varepsilon_1\varepsilon_2 k.$$

It is easy to see that

$$(2.29)\quad \omega_1{}^3 \wedge \omega_2{}^3 + \omega_1{}^4 \wedge \omega_2{}^4 = K\omega^1 \wedge \omega^2, \quad \omega_1{}^3 \wedge \omega_1{}^4 + \omega_2{}^3 \wedge \omega_2{}^4 = k\omega^1 \wedge \omega^2$$

and K is the Gauss' curvature of M.

Definition 2.2. *Let $M: \mathscr{D} \to E^5$ be a surface with* $\dim T_m{}^2(M) = 4$ *for each $m \in M$. The tangent vector field V on M is called* conjugate *if there is a non-trivial tangent vector field V' such that $V'V \subset T(M)$.*

Let us determine the conjugate vector fields on M. Write

$$(2.30)\quad V = \xi v_1 + \eta v_2, \qquad V' = \xi' v_1 + \eta' v_2.$$

Because of

$$(2.31)\quad \begin{aligned} \mathrm{d}V &= (\mathrm{d}x - y\omega_1{}^2)\, v_1 + (\mathrm{d}y + x\omega_1{}^2)\, v_2 + (x\omega_1{}^3 + y\omega_2{}^3)\, v_3 \\ &\quad + (x\omega_1{}^4 + y\omega_2{}^4)\, v_4, \end{aligned}$$

we get

$$(2.32)\quad \begin{aligned} V'V &= \{V'x - y\omega_1{}^2(V')\}\, v_1 + \{V'y + x\omega_1{}^2(V')\}\, v_2 \\ &\quad + (a_1\xi\xi' + a_2\xi\eta' + a_2\eta\xi' + a_3\eta\eta')\, v_3 \\ &\quad + (b_1\xi\xi' + b_2\xi\eta' + b_2\eta\xi' + b_3\eta\eta')\, v_4. \end{aligned}$$

Because of $V'V \subset T(M)$ and $\xi'^2 + \eta'^2 \neq 0$, we get a necessary and sufficient condition for V to be conjugate in the form

$$(2.33)\quad (a_1b_2 - a_2b_1)\,\xi^2 + (a_1b_3 - a_3b_1)\,\xi\eta + (a_2b_3 - a_3b_2)\,\eta^2 = 0.$$

It may be proved that a surface $M: \mathscr{D} \to E^5$ with dim $T_m{}^2(M) = 4$ admits locally non-trivial second order deformations. Therefore, the following theorem has a proper signification.

Theorem 2.2. *Let $M: \mathscr{D} \to E^5$, $\mathscr{D} \subset \mathbf{R}^2$ a bounded domain, be a surface such that:* (i) dim $T_m{}^2(M) = 4$ *for each* $m \in M$; (ii) *there is* $K^2 \neq k^2$ *on* M; (iii) *there are no non-trivial conjugate tangent vector fields on* M. *Let Φ be an infinitesimal second order deformation of M which is trivial on ∂M. Then Φ is trivial on M.*

Proof. From (2.16) and (i), $x_1 = x_2 = 0$, i.e.,

$$(2.34)\quad \varphi_3{}^4 = 0.$$

The equations (2.16), (2.17) and (2.20) reduce to

$$(2.35)\quad \omega_1{}^3 \wedge \varphi_3{}^5 + \omega_1{}^4 \wedge \varphi_4{}^5 = 0,\quad \omega_2{}^3 \wedge \varphi_3{}^5 + \omega_2{}^4 \wedge \varphi_4{}^5 = 0,$$

$$(2.36)\quad \mathrm{d}\varphi_3{}^5 = \omega_3{}^4 \wedge \varphi_4{}^5,\quad \mathrm{d}\varphi_4{}^5 = -\omega_3{}^4 \wedge \varphi_3{}^5,$$

$$(2.37)\quad a_2y_1 - a_1y_2 + b_2z_1 - b_1z_2 = 0,\quad a_3y_1 - a_2y_2 + b_3z_1 - b_2z_2 = 0.$$

From $(2.19_{1,2})$ and (2.36),

$$(2.38)\quad \begin{aligned} &(\mathrm{d}y_1 - y_2\omega_1{}^2 - z_1\omega_3{}^4) \wedge \omega^1 + (\mathrm{d}y_2 + y_1\omega_1{}^2 - z_2\omega_3{}^4) \wedge \omega^2 = 0,\\ &(\mathrm{d}z_1 - z_2\omega_1{}^2 + y_1\omega_3{}^4) \wedge \omega^1 + (\mathrm{d}z_2 + z_1\omega_1{}^2 + y_2\omega_3{}^4) \wedge \omega^2 = 0; \end{aligned}$$

from $(2.1_{1-4}) + (2.23)$,

$$(2.39)\quad \begin{aligned} &(\mathrm{d}a_1 - 2a_2\omega_1{}^2 - b_1\omega_3{}^4) \wedge \omega^1 + \{\mathrm{d}a_2 + (a_1 - a_3)\,\omega_1{}^2 - b_2\omega_3{}^4\} \wedge \omega^2 = 0,\\ &\{\mathrm{d}a_2 + (a_1 - a_3)\,\omega_1{}^2 - b_2\omega_3{}^4\} \wedge \omega^1 + (\mathrm{d}a_3 + 2a_2\omega_1{}^2 - b_3\omega_3{}^4) \wedge \omega^2 = 0,\\ &(\mathrm{d}b_1 - 2b_2\omega_1{}^2 + a_1\omega_3{}^4) \wedge \omega^1 + \{\mathrm{d}b_2 + (b_1 - b_3)\,\omega_1{}^2 + a_2\omega_3{}^4\} \wedge \omega^2 = 0,\\ &\{\mathrm{d}b_2 + (b_1 - b_3)\,\omega_1{}^2 + a_2\omega_3{}^4\} \wedge \omega^1 + (\mathrm{d}b_3 + 2b_2\omega_1{}^2 + a_3\omega_3{}^4) \wedge \omega^2 = 0. \end{aligned}$$

Thus there are functions $S_1, \ldots, S_6, \alpha_1, \ldots, \alpha_4, \beta_1, \ldots, \beta_4$ such that

$$(2.40)\quad \begin{aligned} \mathrm{d}y_1 - y_2\omega_1{}^2 - z_1\omega_3{}^4 &= S_1\omega^1 + S_2\omega^2,\\ \mathrm{d}y_2 + y_1\omega_1{}^2 - z_2\omega_3{}^4 &= S_2\omega^1 + S_3\omega^2,\\ \mathrm{d}z_1 - z_2\omega_1{}^2 + y_1\omega_3{}^4 &= S_4\omega^1 + S_5\omega^2,\\ \mathrm{d}z_2 + z_1\omega_1{}^2 + y_2\omega_3{}^4 &= S_5\omega^1 + S_6\omega^2, \end{aligned}$$

$$(2.41)\quad \begin{aligned} \mathrm{d}a_1 - 2a_2\omega_1{}^2 - b_1\omega_3{}^4 &= \alpha_1\omega^1 + \alpha_2\omega^2,\\ \mathrm{d}a_2 + (a_1 - a_3)\,\omega_1{}^2 - b_2\omega_3{}^4 &= \alpha_2\omega^1 + \alpha_3\omega^2,\\ \mathrm{d}a_3 + 2a_2\omega_1{}^2 - b_3\omega_3{}^4 &= \alpha_3\omega^1 + \alpha_4\omega^2,\\ \mathrm{d}b_1 - 2b_2\omega_1{}^2 + a_1\omega_3{}^4 &= \beta_1\omega^1 + \beta_2\omega^2,\\ \mathrm{d}b_2 + (b_1 - b_3)\,\omega_1{}^2 + a_2\omega_3{}^4 &= \beta_2\omega^1 + \beta_3\omega^2,\\ \mathrm{d}b_3 + 2b_2\omega_1{}^2 + a_3\omega_3{}^4 &= \beta_3\omega^1 + \beta_4\omega^2. \end{aligned}$$

The differential consequences of (2.37) are

$$
\begin{aligned}
(2.42)\quad & a_2S_1 - a_1S_2 + b_2S_4 - b_1S_5 = -\alpha_2y_1 + \alpha_1y_2 - \beta_2z_1 + \beta_1z_2,\\
& a_2S_2 - a_1S_3 + b_2S_5 - b_1S_6 = -\alpha_3y_1 + \alpha_2y_2 - \beta_3z_1 + \beta_2z_2,\\
& a_3S_1 - a_2S_2 + b_3S_4 - b_2S_5 = -\alpha_3y_1 + \alpha_2y_2 - \beta_3z_1 + \beta_2z_2,\\
& a_3S_2 - a_2S_3 + b_3S_5 - b_2S_6 = -\alpha_4y_1 + \alpha_3y_2 - \beta_4z_1 + \beta_3z_2.
\end{aligned}
$$

It is easy to see that the following equations are convenable combinations of the equations (2.37):

$$
\begin{aligned}
(2.43)\quad & (K-k)\,y_1 = (a_2b_2 - a_1b_3)(y_2 + z_1) + (a_1b_2 - a_2b_1 - b_1b_3 + b_2^2)(z_2 - y_1),\\
& (K-k)\,y_2 = (b_1b_3 - b_2^2 + a_3b_2 - a_2b_3)(y_2 + z_1) + (a_2b_2 - a_3b_1)(z_2 - y_1),\\
& (K-k)\,z_1 = (a_1a_3 - a_2^2 - a_1b_2 + a_2b_1)(y_2 + z_1) + (a_3b_1 - a_2b_2)(z_2 - y_1),\\
& (K-k)\,z_2 = (a_2b_2 - a_1b_3)(y_2 + z_1) + (a_1a_3 - a_2^2 + a_3b_2 - a_2b_3)(z_2 - y_1).
\end{aligned}
$$

Similarly, convenable combinations of $(2.42_{1,3})$ lead to

$$
\begin{aligned}
(2.44)\quad (K-k)\,S_2 = {} & (b_1b_3 - b_2^2 + a_3b_2 - a_2b_3)(S_2 + S_4)\\
& + (a_2b_2 - a_3b_1)(S_5 - S_1)\\
& + (a_3 - b_2)(\alpha_2y_1 - \alpha_1y_2 + \beta_2z_1 - \beta_1z_2)\\
& + (b_1 - a_2)(\alpha_3y_1 - \alpha_2y_2 + \beta_3z_1 - \beta_2z_2),\\
(K-k)\,S_5 = {} & (a_2b_2 - a_1b_3)(S_2 + S_4) + (a_1a_3 - a_2^2 + a_3b_2 - a_2b_3)(S_5 - S_1)\\
& + (a_2 + b_3)(\alpha_2y_1 - \alpha_1y_2 + \beta_2z_1 - \beta_1z_2)\\
& - (a_1 + b_2)(\alpha_3y_1 - \alpha_2y_2 + \beta_3z_1 - \beta_2z_2);
\end{aligned}
$$

from $(2.42_{2,4})$, we get

$$
\begin{aligned}
(2.45)\quad (K-k)\,S_2 = {} & (a_2b_2 - a_1b_3)(S_3 + S_5) + (a_1b_2 - a_2b_1 + b_2^2 - b_1b_3)(S_6 - S_2)\\
& + (a_2 + b_3)(\alpha_3y_1 - \alpha_2y_2 + \beta_3z_1 - \beta_2z_2)\\
& - (a_1 + b_2)(\alpha_4y_1 - \alpha_3y_2 + \beta_4z_1 - \beta_3z_2),\\
(K-k)\,S_5 = {} & (a_1a_3 - a_2^2 + a_2b_1 - a_1b_2)(S_3 + S_5) + (a_3b_1 - a_2b_2)(S_6 - S_2)\\
& + (b_2 - a_3)(\alpha_3y_1 - \alpha_2y_2 + \beta_3z_1 - \beta_2z_2)\\
& + (a_2 - b_1)(\alpha_4y_1 - \alpha_3y_2 + \beta_4z_1 - \beta_3z_2).
\end{aligned}
$$

From (2.40),

$$
\begin{aligned}
(2.46)\quad & \mathrm{d}(y_1 - z_2) - (y_2 + z_1)(\omega_1^2 + \omega_3^4) = (S_1 - S_5)\,\omega^1 + (S_2 - S_6)\,\omega^2,\\
& \mathrm{d}(y_2 + z_1) + (y_1 - z_2)(\omega_1^2 + \omega_3^4) = (S_2 + S_4)\,\omega^1 + (S_3 + S_5)\,\omega^2.
\end{aligned}
$$

Let $\mathscr{D}$ be covered by isothermic coordinates (u, v) and let the frames be chosen in such a way that

$$(2.47)\quad \omega^1 = r\,\mathrm{d}u,\quad \omega^2 = r\,\mathrm{d}v;\quad r = r(u, v) > 0.$$

Then

$$(2.48)\qquad \frac{\partial(y_1 - z_2)}{\partial u} = (S_1 - S_5)\, r + \varrho_1(y_2 + z_1),$$

$$\frac{\partial(y_1 - z_2)}{\partial v} = (S_2 - S_6)\, r + \varrho_2(y_2 + z_1),$$

$$\frac{\partial(y_2 + z_1)}{\partial u} = (S_2 + S_4)\, r + \varrho_3(y_1 - z_2),$$

$$\frac{\partial(y_2 + z_1)}{\partial v} = (S_3 + S_5)\, r + \varrho_4(y_1 - z_2),$$

$\varrho_1, \ldots, \varrho_4$ being easy to calculate.

Let us suppose, in accord with (ii),

$$(2.49)\qquad K - k \neq 0.$$

From (2.48), (2.44) and (2.45), we get the system

$$(2.50)\qquad (a_2b_2 - a_3b_1)\frac{\partial(y_1 - z_2)}{\partial u} + (b_1b_3 - b_2^2 + a_2b_1 - a_1b_2)\frac{\partial(y_1 - z_2)}{\partial v}$$

$$+ (b_2^2 - b_1b_3 + a_2b_3 - a_3b_2)\frac{\partial(y_2 + z_1)}{\partial u} + (a_2b_2 - a_1b_3)\frac{\partial(y_2 + z_1)}{\partial v}$$

$$= c_{11}(y_1 - z_2) + c_{12}(y_2 + z_1),$$

$$(a_1a_3 - a_2^2 + a_3b_2 - a_2b_3)\frac{\partial(y_1 - z_2)}{\partial u} + (a_2b_2 - a_3b_1)\frac{\partial(y_1 - z_2)}{\partial v}$$

$$+ (a_1b_3 - a_2b_2)\frac{\partial(y_2 + z_1)}{\partial u} + (a_1a_3 - a_2^2 + a_2b_1 - a_1b_2)\frac{\partial(y_2 + z_1)}{\partial v}$$

$$= c_{21}(y_1 - z_2) + c_{22}(y_2 + z_1).$$

The associated form is equal to

$$(2.51)\qquad (k - K)\,\{(a_1b_2 - a_2b_1)\,\mu^2 + (a_1b_3 - a_3b_1)\,\mu\nu + (a_2b_3 - a_3b_2)\,\nu^2\};$$

it is definite because of the suppositions of our theorem. Thus $y_1 - z_2 = y_2 + z_1 = 0$ inside of $\mathscr{D}$, and (2.43) imply $y_1 = y_2 = z_1 = z_2 = 0$, i.e., $\varphi_3{}^5 = \varphi_4{}^5 = 0$ in $\mathscr{D}$.

In the case $K + k \neq 0$, we have to use $y_1 + z_2$, $y_2 - z_1$ instead of $y_1 - z_2$, $y_2 + z_1$ resp. QED.

VIII. Global differential geometry of hypersurfaces in E^{n+1}

1. An integral formula

By a *hypersurface* $M \subset E^{n+1}$ ($n \geq 2$) we shall mean a mapping $M: \mathscr{G} \to E^{n+1}$, $\mathscr{G}$ being a bounded domain in $\mathbf{R}^n$ with the boundary $\partial\mathscr{G}$ and $(\mathrm{d}M)_g$ injective for each $g \in \mathscr{G}$. As we are going to use the integral formulas and the maximum principle, $\mathscr{G}$ may be well replaced by an n-dimensional manifold with a boundary in the way followed by us in the previous chapters.

To each point $m \in M$, let us associate an orthonormal frame $\{m, v_1, \ldots, v_n, v_{n+1}\}$ such that $v_1, \ldots, v_n \in T_m(M)$. Then we have the fundamental equations

$$(1.1) \qquad \mathrm{d}m = \sum_{i=1}^{n} \omega^i v_i,$$

$$\mathrm{d}v_i = \sum_{j=1}^{n} \omega_i^{\ j} v_j + \omega_i^{\ n+1} v_{n+1}, \qquad \omega_i^{\ j} + \omega_j^{\ i} = 0,$$

$$\mathrm{d}v_{n+1} = -\sum_{i=1}^{n} \omega_i^{\ n+1} v_i$$

with the integrability conditions

$$(1.2) \qquad \sum_{i=1}^{n} \omega^i \wedge \omega_i^{\ n+1} = 0, \qquad \mathrm{d}\omega^i = \sum_{j=1}^{n} \omega^j \wedge \omega_j^{\ i},$$

$$\mathrm{d}\omega_i^{\ j} = \sum_{k=1}^{n} \omega_i^{\ k} \wedge \omega_k^{\ j} - \omega_i^{\ n+1} \wedge \omega_j^{\ n+1},$$

$$\mathrm{d}\omega_i^{\ n+1} = \sum_{j=1}^{n+1} \omega_i^{\ j} \wedge \omega_j^{\ n+1}.$$

From (1.2), we get the existence of functions a_{ij} such that

$$(1.3) \qquad \omega_i^{\ n+1} = \sum_{j=1}^{n} a_{ij} \omega^j, \qquad a_{ij} = a_{ji}.$$

The integrability conditions of (1.3) being

$$(1.4) \qquad \sum_{j=1}^{n} \left(\mathrm{d}a_{ij} - \sum_{k=1}^{n} a_{ik} \omega_j^{\ k} - \sum_{k=1}^{n} a_{kj} \omega_i^{\ k} \right) \wedge \omega^j = 0,$$

we get the existence of functions a_{ijk} such that

$$(1.5) \qquad \mathrm{d}a_{ij} - \sum_{k=1}^{n} a_{ik} \omega_j^{\ k} - \sum_{k=1}^{n} a_{kj} \omega_i^{\ k} = \sum_{k=1}^{n} a_{ijk} \omega^k, \qquad a_{ijk} = a_{jik} = a_{ikj}.$$

Finally, the integrability conditions of (1.5) being

$$\sum_{k=1}^{n}\left(\mathrm{d}a_{ijk}-\sum_{l=1}^{n}a_{ijl}\omega_k{}^l-\sum_{l=1}^{n}a_{ilk}\omega_j{}^l-\sum_{l=1}^{n}a_{ljk}\omega_i{}^l\right)\wedge\omega^k \tag{1.6}$$

$$=\sum_{k,l,m=1}^{n}(a_{ik}a_{jl}+a_{il}a_{jk})\,a_{km}\omega^l\wedge\omega^m,$$

we get the existence of functions a_{ijkl} such that

$$\mathrm{d}a_{ijk}-\sum_{l=1}^{n}a_{ijl}\omega_k{}^l-\sum_{l=1}^{n}a_{ilk}\omega_j{}^l-\sum_{l=1}^{n}a_{ljk}\omega_i{}^l=a_{ijkl}\omega^l, \tag{1.7}$$

$$a_{ijkl}=a_{jikl}=a_{ikjl}, \tag{1.8}$$

$$a_{ijkl}-a_{ijlk}=\sum_{m=1}^{n}(a_{im}a_{jl}+a_{il}a_{jm})\,a_{mk}-\sum_{m=1}^{n}(a_{im}a_{jk}+a_{ik}a_{jm})\,a_{ml}.$$

The first two fundamental forms of M are given by

$$I=\sum_{i=1}^{n}(\omega^i)^2,\qquad II=\sum_{i=1}^{n}\omega^i\omega_i{}^{n+1}=\sum_{i,j=1}^{n}a_{ij}\omega^i\omega^j; \tag{1.9}$$

because of $I=\langle\mathrm{d}m,\mathrm{d}m\rangle$, $II=-\langle\mathrm{d}m,\mathrm{d}v_{n+1}\rangle$, the first one is invariant and the second one is given up to the sign. Without entering into the details (of the elementary theory of quadratic forms), in a neighbourhood of a fixed point $m_0\in M$ we are in the position to choose the frames in such a way that

$$a_{ij}(m_0)=0\quad\text{for}\quad i\neq j;\quad i,j=1,\ldots,n. \tag{1.10}$$

The numbers

$$k_i=a_{ii}(m_0) \tag{1.11}$$

are then the *principal curvatures* of M at m_0. The point $m_0\in M$ is called umbilical if $k_1=\cdots=k_n$; if the hypersurface consists of umbilical points, it is a part of a hypersphere (if $k_1=\cdots=k_n\neq 0$) or a hyperplane (if $k_1=\cdots=k_n=0$).

In what follows, let us restrict ourselves to the hypersurfaces possessing an orthogonal system of lines of curvature, i.e., to hypersurfaces which possess a field of tangent orthonormal frames $\{m, v_1, \ldots, v_n, v_{n+1}\}$ such that

$$a_{ij}=0\quad\text{for}\quad i\neq j,\quad k_i=a_{ii}. \tag{1.12}$$

The equations (1.5) reduce to

$$(k_i-k_j)\,\omega_i{}^j=\sum_{k=1}^{n}a_{ijk}\omega^k\quad\text{for}\quad i\neq j, \tag{1.13}$$

$$\mathrm{d}k_i=\sum_{j=1}^{n}a_{iij}\omega^j,$$

the equations (1.8_2) to

$$a_{ijkl}-a_{ijlk}=0\quad\text{for}\quad (i,j)\neq(k,l)\neq(j,i), \tag{1.14}$$

$$a_{ijij}-a_{ijji}=k_ik_j(k_i-k_j).$$

The following lemma as well the next theorem are the main results of this paragraph.

Lemma 1.1. *For the form*

$$\varphi = \sum_{\substack{i,j=1,\dots,n,\\ i<j}} (-1)^{i+j}(k_i - k_j)^2\, \omega_i{}^j \wedge \omega^1 \cdots \wedge \omega^{i-1} \wedge \omega^{i+1} \wedge \cdots \wedge \omega^{j-1} \wedge \omega^{j+1} \wedge \cdots \wedge \omega^n, \tag{1.15}$$

we have

$$\mathrm{d}\varphi = \left[2 \sum_{i=1}^{n} \left\{ \sum_{\substack{j=1,\dots,n\\ j\neq i}} a_{jji}(a_{jji} - a_{iii}) - \frac{1}{2} \sum_{\substack{j,k=1,\dots,n\\ i\neq j\neq k\neq i}} a_{jji}a_{kki} \right\} + 6 \sum_{\substack{i,j,k=1,\dots,n\\ i<j<k}} a_{ijk}^2 + \sum_{\substack{i,j=1,\dots,n\\ i<j}} (k_i - k_j)^2\, k_i k_j \right] \omega^1 \wedge \cdots \wedge \omega^n. \tag{1.16}$$

The proof is a very tiresome exercise (containing no fun). For $n = 3$, it is not so complicated; nevertheless, let us leave it to the (always kind) reader.

Theorem 1.1. *Let the hypersurface $M \subset E^{n+1}$ with an orthogonal system of lines of curvature satisfy:* (i) $k_i > 0$ *for* $i = 1, \dots, n$; (ii) *the points of* ∂M *are umbilical;* (iii) *there is a function* $f(x_1, \dots, x_n)$ *defined on* $\mathscr{D} \supset k_1(M) \times \cdots \times k_n(M) \subset \mathbf{R}^n$ *such that*

$$f(k_1, \dots, k_n) = 0 \quad \text{on } M, \tag{1.17}$$

$$\frac{\partial f}{\partial x_i} > 0 \quad \text{on } \mathscr{D} \tag{1.18}$$

and the quadratic forms

$$\varkappa_i = \sum_{\substack{j=1,\dots,n,\\ j\neq i}} \left(\frac{\partial f}{\partial x_j} + \frac{\partial f}{\partial x_i} \right) z_j{}^2 + \frac{1}{2} \sum_{\substack{j,k=1,\dots,n\\ i\neq j\neq k\neq i}} \left(\frac{\partial f}{\partial x_j} + \frac{\partial f}{\partial x_k} - \frac{\partial f}{\partial x_i} \right) z_j z_k; \qquad i = 1, \dots, n; \tag{1.19}$$

are positively semi-definite on the subset $\mathscr{D}' \subset \mathscr{D}$ *given by* $f(x_1, \dots, x_n) = 0$. *Then M is a part of a hypersphere.*

Proof. From (1.17) and (1.13), we get

$$\sum_{j=1}^{n} \frac{\partial f}{\partial x_j} a_{jji} = \sum_{\substack{j=1,\dots,n\\ j\neq i}} \frac{\partial f}{\partial x_j} a_{jji} + \frac{\partial f}{\partial x_i} a_{iii}; \qquad i = 1, \dots, n. \tag{1.20}$$

Thus

$$\frac{\partial f}{\partial x_i} \left\{ \sum_{\substack{j=1,\dots,n\\ j\neq i}} a_{jji}(a_{jji} - a_{iii}) - \frac{1}{2} \sum_{\substack{j,k=1,\dots,n\\ i\neq j\neq k\neq i}} a_{jji}a_{kki} \right\} \geqq 0; \qquad i = 1, \dots, n; \tag{1.21}$$

because of (1.19) and (iii); in (1.21), we set $z_j = a_{jji}$. The integrands in the first right-hand term of the integral formula $\int_{\partial M} \varphi = \int_M d\varphi$ being non-negative, we get $k_1 = \cdots = k_n$. QED.

Theorem 1.1 is the *most general* consequence of Lemma 1.1. Let us see, just for $n = 3$, how Lemma 1.1 looks like without the supposition of the existence of an orthogonal system of lines of curvature. The proof is left to the reader, too.

Lemma 1.2. *Let $M \subset E^4$ be a hypersurface. Then the form*

$$
\begin{aligned}
(1.22)\qquad \tau = \{ & a_{13}(a_{133} - a_{111} - a_{122}) + a_{23}(a_{233} - a_{112} - a_{222}) \\
& + 2a_{12}a_{123} + (a_{11} - a_{33})\, a_{113} + (a_{22} - a_{33})\, a_{223}\}\, \omega^1 \wedge \omega^2 \\
& + \{a_{12}(a_{111} - a_{122} + a_{133}) + a_{23}(a_{113} - a_{223} + a_{333}) \\
& \quad - 2a_{13}a_{123} + (a_{22} - a_{11})\, a_{112} + (a_{22} - a_{33})\, a_{233}\}\, \omega^1 \wedge \omega^3 \\
& + \{a_{13}(a_{113} - a_{223} - a_{333}) + a_{12}(a_{112} - a_{233} - a_{222}) \\
& \quad + 2a_{23}a_{123} + (a_{22} - a_{11})\, a_{122} + (a_{33} - a_{11})\, a_{133}\}\, \omega^2 \wedge \omega^3
\end{aligned}
$$

is invariant and we have

$$
\begin{aligned}
(1.23)\qquad d\tau = \Big[& 2(a_{133}^2 - a_{122}a_{133} + a_{122}^2 - a_{111}a_{133} - a_{111}a_{122} \\
& + a_{333}^2 - a_{233}a_{112} + a_{112}^2 - a_{222}a_{112} - a_{222}a_{233} \\
& + a_{113}^2 - a_{113}a_{223} + a_{223}^2 - a_{333}a_{113} - a_{333}a_{223} + 3a_{123}^2) \\
& + \Big\{\frac{1}{2}(a_{11} - a_{22})^2 + \frac{1}{2}(a_{11} - a_{33})^2 + \frac{1}{2}(a_{22} - a_{33})^2 \\
& + 3a_{12}^2 + 3a_{13}^2 + 3a_{23}^2\Big\} H_2 + 3H_1H_3 - H_2{}^2\Big]\, \omega^1 \wedge \omega^2 \wedge \omega^3,
\end{aligned}
$$

where

$$
\begin{aligned}
(1.24)\qquad H_1 &= a_{11} + a_{22} + a_{33}, \\
H_2 &= a_{11}a_{22} + a_{11}a_{33} + a_{22}a_{33} - a_{12}^2 - a_{13}^2 - a_{23}^2, \\
H_3 &= a_{11}a_{22}a_{33} + 2a_{12}a_{13}a_{23} - a_{11}a_{23}^2 - a_{22}a_{13}^2 - a_{33}a_{12}^2
\end{aligned}
$$

are the curvatures of M.

Our next agenda will be to produce consequences of Theorem 1.1 and Lemma 1.2. I restrict myself to the cases $n = 4$ and $n = 3$.

Theorem 1.2. *Let $M \subset E^5$ be a hypersurface with an orthogonal system of lines of curvature. Its curvatures be defined by*

$$(1.25)\qquad H_i = \sum_{1 \leq j_1 < \cdots < j_i \leq 4} k_{j_1} \cdots k_{j_i}; \qquad i = 1, \ldots, 4.$$

Suppose: (i) $k_i > 0$ *for* $i = 1, \ldots, 4$; (ii) *the points on ∂M are umbilical*; (iii) *there is a function $G(y_1, \ldots, y_4)$ defined on $\mathscr{D} = H_1(M) \times \cdots \times H_4(M) \subset \mathbf{R}^4$ such that*

$$(1.26)\qquad G(H_1, \ldots, H_4) = 0 \quad on \quad M$$

and

$$\frac{\partial G}{\partial y_1} \geqq \frac{1}{2} y_1^2 \frac{\partial G}{\partial y_3} + \frac{1}{12} y_1^3 \frac{\partial G}{\partial y_4}, \qquad \frac{\partial G}{\partial y_2} \geqq 0, \qquad \frac{\partial G}{\partial y_3} \geqq 0, \qquad \frac{\partial G}{\partial y_4} \geqq 0 \tag{1.27}$$

on the subset $\mathscr{D}' \subset \mathscr{D}$ *given by* $G(y_1, \ldots, y_n) = 0$; *further, at least one of the inequalities* (1.27) *has to be strict. Then* M *is a part of a hypersphere.*

Proof. (1) Set

$$f^{(1)}(k_1, \ldots, k_4) = H_1, \quad \text{i.e.,} \quad f^{(1)} = x_1 + x_2 + x_3 + x_4. \tag{1.28}$$

Then the associated form $(1.19)_{i=1}$ is

$$\begin{aligned} \varkappa_1^{(1)} &= 2z_2^2 + 2z_3^2 + 2z_4^2 + z_2z_3 + z_2z_4 + z_3z_4 \\ &= \frac{3}{2}(z_2^2 + z_3^2 + z_4^2) + \frac{1}{2}(z_2 + z_3 + z_4)^2 \geqq 0. \end{aligned} \tag{1.29}$$

Because of the symmetry, we get $\varkappa_i^{(1)} \geqq 0$ for $i = 2, 3, 4$.

(2) Set

$$f^{(2)}(k_1, \ldots, k_4) = H_2, \quad \text{i.e.,} \quad f^{(2)} = x_1x_2 + x_1x_3 + x_1x_4 + x_2x_3 + x_2x_4 + x_3x_4. \tag{1.30}$$

Then

$$\begin{aligned} \varkappa_1^{(2)} &= (x_1 + x_2 + 2x_3 + 2x_4)\, z_2^2 + (x_1 + 2x_2 + x_3 + 2x_4)\, z_3^2 \\ &\quad + (x_1 + 2x_2 + 2x_3 + x_4)\, z_4^2 + (2x_1 + x_4)\, z_2z_3 \\ &\quad + (2x_1 + x_3)\, z_2z_4 + (2x_1 + x_2)\, z_3z_4 \\ &= x_1(z_2 + z_3 + z_4)^2 + \frac{1}{2} x_4(z_2 + z_3)^2 + \frac{1}{2} x_3(z_2 + z_4)^2 + \frac{1}{2} x_2(z_3 + z_4)^2 \\ &\quad + \frac{1}{2}(2x_2 + 3x_3 + 3x_4)\, z_2^2 + \frac{1}{2}(3x_2 + 2x_3 + 3x_3)\, z_3^2 \\ &\quad + \frac{1}{2}(3x_2 + 3x_3 + 2x_4)\, z_4^2, \end{aligned} \tag{1.31}$$

and $x_i > 0$ imply $\varkappa_1^{(2)} \geqq 0$. Similarly, we get $\varkappa_i^{(2)} \geqq 0$ for $i = 2, 3, 4$.

(3) Set

$$f^{(3)}(k_1, \ldots, k_4) = H_3 + \frac{1}{6} H_1^3, \tag{1.32}$$

i.e.,

$$f^{(3)} = x_1x_2x_3 + x_1x_2x_4 + x_1x_3x_4 + x_2x_3x_4 + \frac{1}{6}(x_1 + x_2 + x_3 + x_4)^3.$$

Then

$$\begin{aligned} \varkappa_1^{(3)} &= (x_1^2 + x_2^2 + x_3^2 + x_4^2 + 2x_1x_2 + 3x_1x_3 + 3x_1x_4 + 3x_2x_3 + 3x_2x_4 \\ &\quad + 4x_3x_4)\, z_2^2 \\ &\quad + (x_1^2 + x_2^2 + x_3^2 + x_4^2 + 3x_1x_2 + 2x_1x_3 + 3x_1x_4 + 3x_2x_3 \\ &\quad + 4x_2x_4 + 3x_3x_4)\, z_3^2 \end{aligned} \tag{1.33}$$

$$+ (x_1^2 + x_2^2 + x_3^2 + x_4^2 + 3x_1x_2 + 3x_1x_3 + 2x_1x_4 + 4x_2x_3 + 3x_2x_4 + 3x_3x_4)\, z_4^2$$

$$+ \frac{1}{2}(x_1^2 + x_2^2 + x_3^2 + x_4^2 + 4x_1x_2 + 4x_1x_3 + 6x_1x_4 + 2x_2x_4 + 2x_3x_4)\, z_2z_3$$

$$+ \frac{1}{2}(x_1^2 + x_2^3 + x_3^2 + x_4^2 + 4x_1x_2 + 6x_1x_3 + 4x_1x_4 + 2x_2x_3 + 2x_3x_4)\, z_2z_4$$

$$+ \frac{1}{2}(x_1^2 + x_2^2 + x_3^2 + x_4^2 + 6x_1x_2 + 4x_1x_3 + 4x_1x_4 + 2x_2x_3 + 2x_2x_4)\, z_3z_4$$

$$= \frac{1}{2} x_2x_3(z_2 - z_3)^2 + \frac{1}{2} x_2x_4(z_2 - z_4)^2 + \frac{1}{2} x_3x_4(z_3 - z_4)^2$$

$$+ \frac{1}{4}(x_1^2 + x_2^2 + x_3^2 + x_4^2 + 4x_1x_2 + 4x_1x_3 + 6x_1x_4 + 2x_2x_3 + 2x_2x_4 + 2x_3x_4)(z_2 + z_3)^2$$

$$+ \frac{1}{4}(x_1^2 + x_2^2 + x_3^2 + x_4^2 + 4x_1x_2 + 6x_1x_3 + 4x_1x_4 + 2x_2x_3 + 2x_2x_4 + 2x_3x_4)(z_2 + z_4)^2$$

$$+ \frac{1}{4}(x_1^2 + x_2^2 + x_3^2 + x_4^2 + 6x_1x_2 + 4x_1x_3 + 4x_1x_4 + 2x_2x_3 + 2x_2x_4 + 2x_3x_4)(z_3 + z_4)^2$$

$$+ \frac{1}{2}(x_1^2 + x_2^2 + x_3^2 + x_4^2 + x_1x_3 + x_1x_4 + 3x_2x_3 + 3x_2x_4 + 6x_3x_4)\, z_2^2$$

$$+ \frac{1}{2}(x_1^2 + x_2^2 + x_3^2 + x_4^2 + x_1x_2 + x_1x_4 + 3x_2x_3 + 6x_2x_4 + 3x_3x_4)\, z_3^2$$

$$+ \frac{1}{2}(x_1^2 + x_2^2 + x_3^2 + x_4^2 + x_1x_2 + x_1x_3 + 6x_2x_3 + 3x_2x_4 + 3x_3x_4)\, z_4^2,$$

and $\varkappa_1^{(3)} \geqq 0$ as a consequence of $x_i > 0$. The forms $\varkappa_i^{(3)}$ are to be treated similarly.

(4) Set

$$f^{(4)}(k_1, \ldots, k_4) = H_4 + \frac{1}{48} H_1^4, \tag{1.34}$$

i.e.,

$$f^{(4)} = x_1x_2x_3x_4 + \frac{1}{48}(x_1 + x_2 + x_3 + x_4)^4.$$

Then

$$(1.35)\quad \begin{aligned}\varkappa_1^{(4)} = &\left\{x_3x_4(x_1+x_2)+\frac{1}{6}(x_1+x_2+x_3+x_4)^3\right\}z_2^2\\
&+\left\{x_2x_4(x_1+x_3)+\frac{1}{6}(x_1+x_2+x_3+x_4)^3\right\}z_3^2\\
&+\left\{x_2x_3(x_1+x_4)+\frac{1}{6}(x_1+x_2+x_3+x_4)^3\right\}z_4^2\\
&+\left\{x_4(x_1x_2+x_1x_3-x_2x_3)+\frac{1}{12}(x_1+x_2+x_3+x_4)^3\right\}z_2z_3\\
&+\left\{x_3(x_1x_2+x_1x_4-x_2x_4)+\frac{1}{12}(x_1+x_2+x_3+x_4)^3\right\}z_2z_4\\
&+\left\{x_2(x_1x_4+x_1x_3-x_3x_4)+\frac{1}{12}(x_1+x_2+x_3+x_4)^3\right\}z_3z_4\\
= &\frac{1}{2}x_2x_3x_4\{(z_2-z_3)^2+(z_2-z_4)^2+(z_3-z_4)^2\}\\
&+\frac{1}{24}\{12x_1x_4(x_2+x_3)+(x_1+x_2+x_3+x_4)^2\}(z_2+z_3)^2\\
&+\frac{1}{24}\{12x_1x_3(x_2+x_4)+(x_1+x_2+x_3+x_4)^2\}(z_2+z_4)^2\\
&+\frac{1}{24}\{12x_1x_2(x_3+x_4)+(x_1+x_2+x_3+x_4)^2\}(z_3+z_4)^2\\
&+\frac{1}{12}\{(x_1+x_2+x_3+x_4)^3-6x_1x_2(x_3+x_4)\}z_2^2\\
&+\frac{1}{12}\{(x_1+x_2+x_3+x_4)^3-6x_1x_3(x_2+x_4)\}z_3^2\\
&+\frac{1}{12}\{(x_1+x_2+x_3+x_4)^3-6x_1x_4(x_2+x_3)\}z_4^2;\end{aligned}$$

$x_i > 0$ imply $\varkappa_1^{(4)} \geqq 0$ as well as $\varkappa_i^{(4)} \geqq 0$ for $i = 2, 3, 4$.

(5) Let $F(t_1, t_2, t_3, t_4)$ be a function defined on a suitable domain $\mathscr{D}^* \subset \mathbf{R}^4$, and suppose that

$$(1.36)\quad F\left(H_1, H_2, H_3+\frac{1}{6}H_1^3,\ H_4+\frac{1}{48}H_1^4\right)=0 \quad \text{on} \quad M$$

and

$$(1.37)\quad \frac{\partial F}{\partial t_1}\geqq 0,\quad \frac{\partial F}{\partial t_2}\geqq 0,\quad \frac{\partial F}{\partial t_3}\geqq 0,\quad \frac{\partial F}{\partial t_4}\geqq 0$$

on the subset $\mathscr{D}^{**} \subset \mathscr{D}^*$ given by $F(t_1, t_2, t_3, t_4) = 0$; at least one of the inequalities

(1.37) is to be strict. The function f be defined by

$$(1.38)\quad f(x_1, x_2, x_3, x_4) = F\Big(x_1 + x_2 + x_3 + x_4,\ x_1x_2 + x_1x_3 + x_1x_4 + x_2x_3 + x_2x_4 + x_3x_4,\ x_1x_2x_3 + x_1x_2x_4 + x_1x_3x_4 + x_2x_3x_4 + \frac{1}{6}(x_1 + x_2 + x_3 + x_4)^3,\ x_1x_2x_3x_4 + \frac{1}{48}(x_1 + x_2 + x_3 + x_4)^4\Big).$$

From $x_i > 0$, we get (1.18), and the associated forms (1.19) are given by

$$(1.39)\quad \varkappa_i = \frac{\partial F}{\partial t_1}\varkappa_i^{(1)} + \frac{\partial F}{\partial t_2}\varkappa_i^{(2)} + \frac{\partial F}{\partial t_3}\varkappa_i^{(3)} + \frac{\partial F}{\partial t_4}\varkappa_i^{(4)},$$

$\varkappa_i^{(j)}$ being given by (1.29), (1.31), (1.33), (1.35) and analogous equations resp. Thus $x_i > 0$ imply $\varkappa_i \geqq 0$, and, according to Theorem 1.1, each hypersurface M satisfying (i), (ii) and (1.36) + (1.37) is a part of a hypersphere.

(6) Finally, let us start with the function $G(y_1, \ldots, y_4)$ of our theorem. The function F be defined by

$$(1.40)\quad F(t_1, t_2, t_3, t_4) = G\left(t_1, t_2, t_3 - \frac{1}{6}t_1^3, t_4 - \frac{1}{48}t_1^4\right).$$

Then

$$(1.41)\quad \frac{\partial F}{\partial t_1} = \frac{\partial G}{\partial y_1} - \frac{1}{2}t_1^2\frac{\partial G}{\partial y_3} - \frac{1}{12}t_1^3\frac{\partial G}{\partial y_4},$$
$$\frac{\partial F}{\partial t_2} = \frac{\partial G}{\partial y_2}, \qquad \frac{\partial F}{\partial t_3} = \frac{\partial G}{\partial y_3}, \qquad \frac{\partial F}{\partial t_4} = \frac{\partial G}{\partial y_4}$$

and (1.37) are satisfied because of (1.27). QED.

Theorem 1.3. *Let $M \subset E^4$ be a hypersurface with an orthogonal system of lines of curvature. Suppose:* (i) $k_i > 0$ *for* $i = 1, 2, 3$; (ii) *the boundary ∂M of M consists of umbilical points*; (iii) *there is, on M,*

$$(1.42)\quad A_1\,\mathrm{d}k_1 + A_2\,\mathrm{d}k_2 + A_3\,\mathrm{d}k_3 = 0,$$

$A_i\colon M \to \mathbf{R}$ being functions such that

$$(1.43)\quad 4(A_i + A_j)(A_i + A_k) - (A_j + A_k - A_i)^2 \geqq 0 \quad \text{for } i, j, k = 1, 2, 3;\ i \neq j \neq k \neq i.$$

Then M is a part of a hypersphere.

Proof. For $i \neq j \neq k \neq i$; $i, j, k = 1, 2, 3$;

$$(1.44)\quad \varkappa_i' := \sum_{\substack{r=1,2,3\\ r\neq i}} a_{rri}(a_{rri} - a_{iii}) - \frac{1}{2}\sum_{\substack{r,s=1,2,3\\ i\neq r\neq s\neq i}} a_{rri}a_{ssi}$$
$$= a_{ijj}^2 + a_{ikk}^2 - a_{ijj}a_{ikk} - a_{iii}(a_{ijj} + a_{ikk})$$

From (1.13_2) and (1.42),

$$A_i a_{iii} = -A_j a_{ijj} - A_k a_{ikk} \tag{1.45}$$

and

$$A_i \varkappa_i' = (A_i + A_j)\, a_{ijj}^2 + (A_j + A_k - A_i)\, a_{ijj} a_{ikk} + (A_i + A_k)\, a_{ikk}^2. \tag{1.46}$$

Because of the supposition (1.43), $\varkappa_i' \geqq 0$ for $i = 1, 2, 3$. From the integral formula $\int_{\partial M} \varphi = \int_M d\varphi$, see (1.15) and (1.16), $k_i = k_j$. QED.

Theorem 1.4. *Let $M \subset E^4$ be a hypersurface with an orthogonal system of lines of curvature. Suppose:* (i) $k_i > 0$ *for* $i = 1, 2, 3$; (ii) *the boundary ∂M of M consists of umbilical points*; (iii) *there is a function $F(H_1, H_2, H_3)$ such that, on M,*

$$F(H_1, H_2, H_3) = 0 \tag{1.47}$$

and

$$\frac{\partial F}{\partial H_1} + (k_i + k_j) \frac{\partial F}{\partial H_2} + k_i k_j \frac{\partial F}{\partial H_3} > 0, \tag{1.48}$$

$$\begin{aligned} &15 \left(\frac{\partial F}{\partial H_1}\right)^2 + 4(2k_i^2 + 2k_j^2 + 5k_ik_j + 3k_ik_k + 3k_jk_k) \left(\frac{\partial F}{\partial H_2}\right)^2 \\ &+ (3k_i^2k_j^2 - k_i^2k_k^2 - k_j^2k_k^2 + 6k_i^2k_jk_k + 6k_ik_j^2k_k + 2k_ik_jk_k^2) \left(\frac{\partial F}{\partial H_3}\right)^2 \\ &+ 12(2k_i + 2k_j + k_k) \frac{\partial F}{\partial H_1} \frac{\partial F}{\partial H_2} + 6(3k_ik_j + k_ik_k + k_jk_k) \frac{\partial F}{\partial H_1} \frac{\partial F}{\partial H_3} \\ &+ 4(3k_i^2k_j + k_i^2k_k + 3k_ik_j^2 + k_j^2k_k + 7k_ik_jk_k) \frac{\partial F}{\partial H_2} \frac{\partial F}{\partial H_3} \geqq 0 \end{aligned} \tag{1.49}$$

for $i \neq j \neq k \neq i$; $i, j, k = 1, 2, 3$. *Then M is a part of a hypersphere.* (*Here, the curvatures H_i are defined by* (1.24) *with regard to* (1.12).)

Proof. From (1.47), we get (1.42) with

$$A_i = \frac{\partial F}{\partial H_1} + (k_j + k_k) \frac{\partial F}{\partial H_2} + k_j k_k \frac{\partial F}{\partial H_3}; \quad i \neq j \neq k \neq i; \quad i, j, k = 1, 2, 3; \tag{1.50}$$

and the conditions (1.43) become exactly (1.48) and (1.49). QED.

Theorem 1.5. *Let $M \subset E^4$ be a hypersurface with an orthogonal system of lines of curvature. Suppose:* (i) $k_i > 0$ *for* $i = 1, 2, 3$; (ii) *the boundary ∂M of M consists of umbilical points*; (iii) *we have, on M,*

$$f(H_1, H_2, rH_1H_2 + H_3) = 0, \tag{1.51}$$

$r \in \mathbf{R}$ *satisfying* $83r \geqq 6\sqrt{3} - 5$ *and $f(\alpha, \beta, \gamma)$ being a function with one of its derivatives positive and others non-negative. Then M is a part of a hypersphere.*

Proof. Set

$$F(H_1, H_2, H_3) = f(H_1, H_2, rH_1H_2 + H_3). \tag{1.52}$$

Then

(1.53) $$\frac{\partial F}{\partial H_1} = \frac{\partial f}{\partial \alpha} + rH_2 \frac{\partial f}{\partial \gamma}, \qquad \frac{\partial F}{\partial H_2} = \frac{\partial f}{\partial \beta} + rH_1 \frac{\partial f}{\partial \gamma}, \qquad \frac{\partial F}{\partial H_3} = \frac{\partial f}{\partial \gamma},$$

and the left-hand side of (1.48) is equal to

(1.54) $$\frac{\partial f}{\partial \alpha} + (k_i + k_j) \frac{\partial f}{\partial \beta} + \{r(k_i + k_j)(2H_1 - k_i - k_j) + k_i k_j\} \frac{\partial f}{\partial \gamma},$$

thus positive. Further, the left-hand side of (1.49) is equal to

(1.55) $$\mu_1 \left(\frac{\partial f}{\partial \alpha}\right)^2 + \mu_2 \left(\frac{\partial f}{\partial \beta}\right)^2 + \mu_3 \frac{\partial f}{\partial \alpha} \frac{\partial f}{\partial \beta} + \mu_4 \frac{\partial f}{\partial \alpha} \frac{\partial f}{\partial \gamma}$$
$$+ \mu_5 \frac{\partial f}{\partial \beta} \frac{\partial f}{\partial \gamma} + \{\mu_6 + (83r^2 - 10r - 1) k_k^2(k_i^2 + k_j^2)\} \left(\frac{\partial f}{\partial \gamma}\right)^2$$

with $\mu_A = \mu_A(k_1, k_2, k_3) \geqq 0$ for $k_1 \geqq 0$, $k_2 \geqq 0$, $k_3 \geqq 0$. QED.

Theorem 1.6. *Let $M \subset E^4$ be a hypersurface with an orthogonal system of lines of curvature. Suppose:* (i) $k_i > 0$ *for* $i = 1, 2, 3$; (ii) *the boundary ∂M of M consists of umbilical points*; (iii) *we have, on M,*

(1.56) $$H_3 = \text{const}, \qquad 4H_1H_3 \geqq H_2^2.$$

Then M is a part of a hypersphere.

Proof. For $F(H_1, H_2, H_3) = H_3$, the left-hand side of (1.49) is equal to

(1.57) $$3k_i^2k_j^2 - k_i^2k_k^2 - k_j^2k_k^2 + 6k_i^2k_jk_k + 6k_ik_j^2k_k + 2k_ik_jk_k^2$$
$$= 4H_1H_3 - H_2^2 + 4k_ik_jH_2.$$

Thus (1.56_2) implies (1.49). QED.

Finally, let us present consequences of Lemma 1.2. For (1.12), we get

(1.58) $$\left\{\frac{1}{2}(a_{11} - a_{22})^2 + \frac{1}{2}(a_{11} - a_{33})^2 + \frac{1}{2}(a_{22} - a_{33})^2 + 3a_{12}^2 + 3a_{13}^2 + 3a_{23}^2\right\} H_2$$
$$+ 3H_1H_3 - H_2^2 = (k_1 - k_2)^2 k_1k_2 + (k_1 - k_3)^2 k_1k_3 + (k_2 - k_3)^2 k_2k_3.$$

Thus Theorem 1.5 holds true also without the supposition of the existence of an orthogonal system of lines of curvature. Indeed, consider the differential consequences of (1.52). Then take a fixed (but quite arbitrary) point $m_0 \in M$. We are able to choose the frames in such a way that (1.12) holds true at m_0. Following the proof of Theorem 1.5 at m_0, we get $k_1 = k_2 = k_3$ at m_0; m_0 being arbitrary, we are done.

Finally, let us present a very wide generalization of our integral formulas presented in Lemmas 1.1, 1.2 and IV.3.5.

Let M be a connected orientable n-dimensional manifold M endowed by a quadratic positive definite differential form ds^2; the couple (M, ds^2) is the so-called *Riemannian manifold*. In a suitable domain $U \subset M$, there are 1-forms $\omega^1, \ldots, \omega^n$

such that

$$(1.59) \qquad ds^2 = \delta_{ij}\omega^i\omega^j = \sum_i (\omega^i)^2;$$

recall that $\delta_{ii} = 1$ and $\delta_{ij} = 0$ for $i \neq j$.

Lemma 1.3. *Let* (M, ds^2) *be a Riemannian manifold, and let* (1.59) *take hold in* $U \subset M$. *Then there is, in* U, *a unique set of* 1-*forms* $\omega_i{}^j$ *such that*

$$(1.60) \qquad d\omega^i = \omega^j \wedge \omega_j{}^i, \quad \delta_{ik}\omega_j{}^k + \delta_{jk}\omega_i{}^k = 0.$$

Proof. Let us write, always in U,

$$(1.61) \qquad d\omega^i = A^i_{jk}\omega^j \wedge \omega^k, \quad A^i_{jk} + A^i_{kj} = 0.$$

Let us suppose the existence of 1-forms $\omega_i{}^j$ satisfying (1.60). Then (1.60_1) and (1.61) imply

$$(1.62) \qquad \omega^i \wedge (\omega_i{}^j - A^j_{ik}\omega^k) = 0,$$

and we get the existence of functions B^i_{jk} such that

$$(1.63) \qquad \omega_i{}^j = A^j_{ik}\omega^k + B^j_{ik}\omega^k, \quad B^i_{jk} - B^i_{kj} = 0.$$

Inserting these into (1.60_2), we get

$$(1.64) \qquad \delta_{il}A^l_{jk} + \delta_{jl}A^l_{ik} + \delta_{il}B^l_{jk} + \delta_{jl}B^l_{ik} = 0;$$

the permutation of i, j, k leads to

$$(1.65) \qquad \begin{aligned} &\delta_{il}A^l_{kj} + \delta_{kl}A^l_{ij} + \delta_{il}B^l_{kj} + \delta_{kl}B^l_{ij} = 0, \\ &\delta_{kl}A^l_{ji} + \delta_{jl}A^l_{ki} + \delta_{kl}B^l_{ji} + \delta_{jl}B^l_{ki} = 0. \end{aligned}$$

Subtracting (1.65) from (1.64), we get

$$(1.66) \qquad B^i_{jk} = \delta^{li}\delta_{mj}A^m_{kl} + \delta^{li}\delta_{mk}A^m_{jl},$$

i.e.,

$$(1.67) \qquad \omega_i{}^j = (A^j_{ik} + \delta^{lj}\delta_{mi}A^m_{kl} + \delta^{lj}\delta_{mk}A^m_{jl})\,\omega^k.$$

This proves the unicity. The forms (1.67) satisfying (1.60), we are done. QED.

Now, let us change our coframes $(\omega^1, \ldots, \omega^n)$ into $(\tau^1, \ldots, \tau^n)$, i.e., let us suppose, in U,

$$(1.68) \qquad ds^2 = \delta_{ij}\tau^i\tau^j$$

and

$$(1.69) \qquad \omega^i = T_j{}^i\tau^j.$$

The matrix $\|T_i{}^j\|$ is, of course, orthogonal, i.e.,

$$(1.70) \qquad \delta_{ij}T_k{}^iT_l{}^j = \delta_{kl}.$$

Let us denote by $\|\tilde{T}_i{}^j\|$ the inverse matrix to $\|T_i{}^j\|$, i.e., let

$$(1.71) \qquad T_i{}^k\tilde{T}_k{}^j = \tilde{T}_i{}^kT_k{}^j = \delta_i{}^j;$$

here, $\delta_i^i = 1$ and $\delta_i^j = 0$ for $i \neq j$. Let the matrix $\|\tau_i^j\|$ of 1-forms correspond to the coframes $(\tau^1, \ldots, \tau^n)$, i.e., let

$$d\tau^i = \tau^j \wedge \tau_j^i, \qquad \delta_{ik}\tau_j^k + \delta_{jk}\tau_i^k = 0. \tag{1.72}$$

A direct verification implies

Lemma 1.4. *We have*

$$\tau_i^j = \tilde{T}_k^j(dT_i^k + T_i^l\omega_l^k). \tag{1.73}$$

The components R^j_{ikl} of the *curvature tensor* of (M, ds^2) with respect to the coframes $(\omega^1, \ldots, \omega^n)$ be introduced by

$$d\omega_i^j = \omega_i^k \wedge \omega_k^j - \frac{1}{2} R^j_{ikl}\, \omega^k \wedge \omega^l, \qquad R^j_{ikl} + R^j_{ilk} = 0. \tag{1.74}$$

Lemma 1.5. *Let $\tilde{R}^j_{ikl}$ be the components of the curvature tensor of (M, ds^2) with respect to the coframes $(\tau^1, \ldots, \tau^n)$. Then*

$$\tilde{R}^j_{ikl} = T_i^m T_k^n T_l^p \tilde{T}_q^j R^q_{mnp}. \tag{1.75}$$

Proof. The formula (1.75) follows easily if substituting (1.73) into

$$d\tau_i^j = \tau_i^k \wedge \tau_k^j - \frac{1}{2} \tilde{R}^j_{ikl}\, \tau^k \wedge \tau^l, \qquad \tilde{R}^j_{ikl} + \tilde{R}^j_{ilk} = 0 \tag{1.76}$$

and using (1.71) together with its differential consequence

$$d\tilde{T}_i^j = -\tilde{T}_k^j \tilde{T}_i^l\, dT_l^k. \tag{1.77}$$

QED.

Lemma 1.6. *We have*

$$\begin{aligned} &\delta_{im}R^m_{jkl} + \delta_{jm}R^m_{ikl} = 0, \qquad \delta_{jm}R^m_{ikl} = \delta_{lm}R^m_{kij}, \\ &R^i_{jkl} + R^i_{klj} + R^i_{ljk} = 0. \end{aligned} \tag{1.78}$$

Proof. The identity (1.78_1) follows from (1.60_2) and (1.74). The exterior differentiation of (1.60_1) implies $R^i_{jkl}\omega^j \wedge \omega^k \wedge \omega^l = 0$, i.e., (1.78_3). Further, using (1.74_2) and (1.78_1), (1.78_3) yields

$$\begin{aligned} R^j_{ikl} &= R^k_{jli} + R^l_{jik} = -R^k_{lij} - R^k_{ijl} - R^l_{ikj} - R^l_{kji} \\ &= 2R^l_{kij} + R^i_{kjl} + R^i_{lkj} = 2R^l_{kij} - R^i_{jlk} = 2R^l_{kij} - R^j_{ikl}, \end{aligned}$$

i.e., (1.78_2). QED.

On M, be given a quadratic differential form Q; in U, it may be written as

$$Q = a_{ij}\omega^i\omega^j = \tilde{a}_{ij}\tau^i\tau^j; \qquad a_{ij} = a_{ji}, \quad \tilde{a}_{ij} = \tilde{a}_{ji}. \tag{1.79}$$

We have

$$\tilde{a}_{ij} = T_i^k T_j^l a_{kl}. \tag{1.80}$$

The *covariant derivatives* b_{ijk} of a_{ij} with respect to the coframes $(\omega^1, \ldots, \omega^n)$ be defined by

$$(1.81)\qquad \mathrm{d}a_{ij} - a_{ik}\omega_j^{\ k} - a_{kj}\omega_i^{\ k} = b_{ijk}\omega^k, \qquad b_{ijk} = b_{jik}.$$

Substituting into the analogous equations

$$(1.82)\qquad \mathrm{d}\tilde{a}_{ij} - \tilde{a}_{ik}\tau_j^{\ k} - \tilde{a}_{kj}\tau_i^{\ k} = \tilde{b}_{ijk}\tau^k,$$

we get

$$(1.83)\qquad \tilde{b}_{ijk} = T_i^{\ l}T_j^{\ m}T_k^{\ n}b_{lmn}.$$

The exterior differentiation of (1.81) yields

$$(1.84)\qquad (\mathrm{d}b_{ijk} - b_{ljk}\omega_i^{\ l} - b_{ilk}\omega_j^{\ l} - b_{ijl}\omega_k^{\ l}) \wedge \omega^k = \frac{1}{2}(a_{ik}R^k_{jmn} + a_{kj}R^k_{imn})\,\omega^m \wedge \omega^n$$

and the existence of functions c_{ijkl} such that

$$(1.85)\qquad \begin{aligned} &\mathrm{d}b_{ijk} - b_{ljk}\omega_i^{\ l} - b_{ilk}\omega_j^{\ l} - b_{ijl}\omega_k^{\ l} = c_{ijkl}\omega^l, \\ &c_{ijkl} = c_{jikl}, \qquad c_{ijkl} - c_{ijlk} = -a_{im}R^m_{jkl} - a_{mj}R^m_{ikl}. \end{aligned}$$

On U, let us define the 1-forms

$$(1.86)\qquad \tau_1 = \delta^{ij}\delta^{kl}a_{im}b_{jkl}\omega^m, \qquad \tau_2 = \delta^{ij}\delta^{kl}a_{ik}b_{jml}\omega^m.$$

Because of (1.80), (1.83) and (1.69), *the forms* τ_1 *and* τ_2 *are globally defined over all of* M. The $*$-*operator* on $(M, \mathrm{d}s^2)$ be defined by

$$(1.87)\qquad *\omega^i = (-1)^{i+1}\,\omega^1 \wedge \cdots \wedge \omega^{i-1} \wedge \omega^{i+1} \wedge \cdots \wedge \omega^n,$$

i.e.,

$$\omega^i \wedge *\,\omega^i = \omega^1 \wedge \cdots \wedge \omega^n =: \mathrm{d}o.$$

By a direct calculation, we get

$$(1.88)\qquad \begin{aligned} \mathrm{d} * \tau_1 &= \delta^{ij}\delta^{kl}\delta^{mn}(b_{imm}b_{jkl} + a_{im}c_{jkln})\,\mathrm{d}o, \\ \mathrm{d} * \tau_2 &= \delta^{ij}\delta^{kl}\delta^{mn}(b_{ikm}b_{jnl} + a_{ik}c_{jmln})\,\mathrm{d}o. \end{aligned}$$

Using the relation $(1.85)_3$, we get

$$(1.89)\qquad \mathrm{d} * (\tau_1 - \tau_2) = \{\delta^{ij}\delta^{kl}\delta^{mn}(b_{imn}b_{jkl} - b_{ikm}b_{jln}) - \delta^{ij}a_{ki}a_{lm}(\delta^{kl}\delta^{np} + \delta^{kn}\delta^{lp})\,R^m_{npj}\}\,\mathrm{d}o.$$

Further,

$$(1.90)\qquad \begin{aligned} &\delta^{ij}a_{ki}a_{lm}(\delta^{kl}\delta^{np} + \delta^{kn}\delta^{lp})\,R^m_{npj} \\ &= \sum_{\alpha}(a_{\alpha\alpha})^2 \cdot \sum_{\beta \neq \alpha} R^{\alpha}_{\beta\beta\alpha} - 2\sum_{\alpha \neq \beta} a_{\alpha\alpha}a_{\beta\beta}R^{\beta}_{\alpha\alpha\beta} \\ &\quad + \sum_{\gamma \neq \alpha}\ \sum_{\delta \neq \varepsilon}\ \sum_{\beta,\varphi,\psi} a_{\gamma\alpha}a_{\delta\varepsilon}\delta^{\alpha\beta}(\delta^{\gamma\delta}\delta^{\varphi\psi} + \delta^{\gamma\varphi}\delta^{\delta\psi})\,R^{\varepsilon}_{\varphi\psi\beta}, \end{aligned}$$

$$(1.91)\qquad \tau_1 = \sum_{\alpha,\beta} a_{\alpha\alpha} b_{\alpha\beta\beta}\omega^\alpha + \sum_{\alpha \neq \varepsilon} \sum_{\beta,\gamma,\delta} a_{\alpha\varepsilon}\delta^{\alpha\beta}\delta^{\gamma\delta} b_{\beta\gamma\delta}\omega^\varepsilon,$$

$$\tau_2 = \sum_{\alpha,\beta} a_{\beta\beta} b_{\alpha\beta\beta}\omega^\alpha + \sum_{\alpha \neq \gamma} \sum_{\beta,\delta,\varepsilon} a_{\alpha\gamma}\delta^{\alpha\beta}\delta^{\gamma\delta} b_{\beta\varepsilon\delta}\omega^\varepsilon.$$

Theorem 1.7. *Let* $(M, \mathrm{d}s^2)$ *be a connected orientable n-dimensional Riemannian manifold and* Q *a quadratic differential form on* M. *Let us suppose:* (i) *there is a function* $\lambda\colon \partial M \to \mathbf{R}$ *such that* $Q = \lambda\, \mathrm{d}s^2$ *on the boundary* ∂M *of* M; (ii) *for the invariant*

$$(1.92)\qquad B := \delta^{ij}\delta^{kl}\delta^{mn}(b_{imn}b_{jkl} - b_{ikm}b_{jln}),$$

we have

$$(1.93)\qquad B \leqq 0$$

on M; (iii) *to each point* $m \in M$, *there is a neighbourhood* $U \subset M$ *of* m *and a field of coframes* $(\omega^1, \ldots, \omega^n)$ *on* U *such that*

$$(1.94)\qquad \mathrm{d}s^2 = \sum_i (\omega^i)^2, \qquad Q = \sum_i \lambda_i(\omega^i)^2 \quad \text{at } m$$

and

$$(1.95)\qquad R^j_{iij}(m) > 0 \quad \text{for each} \quad i, j = 1, \ldots, n; \quad i \neq j.$$

Then there is a function $\Lambda\colon M \to \mathbf{R}$ *such that* $Q = \Lambda\, \mathrm{d}s^2$ *on* M.

Proof. We have the integral formula — see (1.89) and (1.90) —

$$(1.96)\qquad \int_{\partial M} * (\tau_1 - \tau_2) = \int_M \Big\{B - \sum_{i<j} (\lambda_i - \lambda_j)^2 R^j_{iij}\Big\}\, \mathrm{d}o.$$

The left-hand side being equal to 0 because of (i) and (1.91), we have $\lambda_1 = \cdots = \lambda_n$ on M. Now, set $\Lambda = \lambda_1$. QED.

The reader is invited to work out the consequences for Q the second fundamental form of a hypersurface M in E^{n+1}; all he has to do is to check that for $II = \sum_i k_i(\omega^i)^2$ at $m \in M$ (in a convenable field of coframes $\omega^1, \ldots, \omega^n$ around m), we have $R^j_{iij} = k_i k_j$ for $i \neq j$, k_i being the principal curvatures of M at m.

2. The maximum principle

In this section, we are going to characterize the hyperspheres of E^4 using the maximum principle as introduced in Theorems III.2.1 and III.2.2.

Let $M \subset E^4$ be a hypersurface; the equations (1.3), (1.5), (1.7) and (1.8) be re-written in the form

$$(2.1)\qquad \omega_1{}^4 = a_{11}\omega^1 + a_{12}\omega^2 + a_{13}\omega^3,$$

$$\omega_2{}^4 = a_{12}\omega^1 + a_{22}\omega^2 + a_{23}\omega^3,$$

$$\omega_3{}^4 = a_{13}\omega^1 + a_{23}\omega^2 + a_{33}\omega^3;$$

$$
\begin{aligned}
(2.2)\qquad & \mathrm{d}a_{11} - 2a_{12}\omega_1{}^2 - 2a_{13}\omega_1{}^3 = \alpha\omega^1 + \beta\omega^2 + \gamma\omega^3,\\
& \mathrm{d}a_{12} + (a_{11} - a_{22})\,\omega_1{}^2 - a_{23}\omega_1{}^3 - a_{13}\omega_2{}^3 = \beta\omega^1 + \lambda\omega^2 + \theta\omega^3,\\
& \mathrm{d}a_{13} - a_{23}\omega_1{}^2 + (a_{11} - a_{33})\,\omega_1{}^3 + a_{12}\omega_2{}^3 = \gamma\omega^1 + \theta\omega^2 + \xi\omega^3,\\
& \mathrm{d}a_{22} + 2a_{12}\omega_1{}^2 - 2a_{23}\omega_2{}^3 = \lambda\omega^1 + \psi\omega^2 + \varrho\omega^3,\\
& \mathrm{d}a_{23} + a_{13}\omega_1{}^2 + a_{12}\omega_1{}^3 + (a_{22} - a_{33})\,\omega_2{}^3 = \theta\omega^1 + \varrho\omega^2 + \sigma\omega^3,\\
& \mathrm{d}a_{33} + 2a_{13}\omega_1{}^3 + 2a_{23}\omega_2{}^3 = \xi\omega^1 + \sigma\omega^2 + \varepsilon\omega^3;
\end{aligned}
$$

$$
\begin{aligned}
(2.3)\qquad & \mathrm{d}\alpha - 3\beta\omega_1{}^2 - 3\gamma\omega_1{}^3 = A\omega^1 + B\omega^2 + C\omega^3,\\
& \mathrm{d}\beta + (\alpha - 2\lambda)\,\omega_1{}^2 - 2\theta\omega_1{}^3 - \gamma\omega_2{}^3 = B'\omega^1 + D\omega^2 + E\omega^3,\\
& \mathrm{d}\gamma - 2\theta\omega_1{}^2 + (\alpha - 2\xi)\,\omega_1{}^3 + \beta\omega_2{}^3 = C'\omega^1 + E'\omega^2 + F\omega^3,\\
& \mathrm{d}\lambda + (2\beta - \psi)\,\omega_1{}^2 - \varrho\omega_1{}^3 - 2\theta\omega_2{}^3 = G\omega^1 + M\omega^2 + N\omega^3,\\
& \mathrm{d}\psi + 3\lambda\omega_1{}^2 - 3\varrho\omega_2{}^3 = M'\omega^1 + P\omega^2 + Q\omega^3,\\
& \mathrm{d}\varrho + 2\theta\omega_1{}^2 + \lambda\omega_1{}^3 - (2\sigma - \psi)\,\omega_2{}^3 = N'\omega^1 + Q'\omega^2 + R\omega^3,\\
& \mathrm{d}\xi - \sigma\omega_1{}^2 + (2\gamma - \varepsilon)\,\omega_1{}^3 + 2\theta\omega_2{}^3 = S\omega^1 + T\omega^2 + U\omega^3,\\
& \mathrm{d}\sigma - \xi\omega_1{}^2 + 2\theta\omega_1{}^3 + (2\varrho - \varepsilon)\,\omega_2{}^3 = T'\omega^1 + V\omega^2 + Z\omega^3,\\
& \mathrm{d}\varepsilon + 3\xi\omega_1{}^3 + 3\sigma\omega_2{}^3 = U'\omega^1 + Z'\omega^2 + W\omega^3,\\
& \mathrm{d}\theta + (\gamma - \varrho)\,\omega_1{}^2 + (\beta - \sigma)\,\omega_1{}^3 + (\lambda - \xi)\,\omega_2{}^3 = X\omega^1 + Y\omega^2 + I\omega^3;
\end{aligned}
$$

$$
\begin{aligned}
(2.4)\qquad & B' - B = 2a_{11}a_{12}a_{22} - 2a_{12}^3 + 2a_{11}a_{13}a_{23} - 2a_{12}a_{13}^2,\\
& E' - E = 2a_{12}^2a_{23} - 2a_{12}a_{13}a_{22} + 2a_{12}a_{13}a_{33} - 2a_{13}^2a_{23},\\
& C' - C = 2a_{11}a_{12}a_{23} - 2a_{12}^2a_{13} + 2a_{11}a_{13}a_{33} - 2a_{13}^3,\\
& M' - M = -2a_{11}a_{12}a_{22} + 2a_{12}^3 + 2a_{12}a_{23}^2 - 2a_{13}a_{22}a_{23},\\
& Q' - Q = -2a_{12}^2a_{23} + 2a_{12}a_{13}a_{22} + 2a_{22}a_{23}a_{33} - 2a_{23}^3,\\
& N' - N = -2a_{11}a_{12}a_{23} + 2a_{12}^2a_{13} + 2a_{12}a_{23}a_{33} - 2a_{13}a_{23}^2,\\
& T' - T = -2a_{11}a_{13}a_{23} + 2a_{12}a_{13}^2 - 2a_{12}a_{23}^2 + 2a_{13}a_{22}a_{23},\\
& Z' - Z = -2a_{12}a_{13}a_{33} + 2a_{13}^2a_{23} - 2a_{22}a_{23}a_{33} + 2a_{23}^2,\\
& U' - U = -2a_{11}a_{13}a_{33} + 2a_{13}^3 - 2a_{12}a_{23}a_{33} + 2a_{13}a_{23}^2,\\
& G - D = -a_{11}^2a_{22} + a_{11}a_{12}^2 + a_{11}a_{22}^2 - a_{22}a_{12}^2 + a_{11}a_{23}^2 - a_{13}^2a_{22},\\
& Y - N = -a_{11}a_{12}a_{23} + a_{11}a_{13}a_{22} + a_{12}a_{22}a_{23} - a_{13}a_{22}^2 + a_{12}a_{23}a_{33}\\
& \qquad\quad - 2a_{13}a_{23}^2 + a_{13}a_{22}a_{33},\\
& X - E = -a_{11}^2a_{23} + a_{11}a_{12}a_{13} + a_{11}a_{22}a_{23} - a_{12}a_{13}a_{22} + a_{11}a_{23}a_{33}\\
& \qquad\quad - 2a_{13}^2a_{23} + a_{12}a_{13}a_{33},\\
& T - I = 2a_{12}a_{23}^2 - a_{13}a_{22}a_{23} - a_{11}a_{12}a_{33} + a_{11}a_{13}a_{23} + a_{12}a_{33}^2\\
& \qquad\quad - a_{13}a_{23}a_{33} - a_{12}a_{22}a_{33},\\
& S - F = a_{11}a_{23}^2 - a_{11}^2a_{33} + a_{11}a_{13}^2 + a_{11}a_{33}^2 - a_{13}^2a_{33} - a_{12}^2a_{33},\\
& V - R = a_{13}^2a_{22} - a_{12}^2a_{33} - a_{22}^2a_{33} + a_{22}a_{23}^2 + a_{22}a_{33}^2 - a_{23}^2a_{33}.
\end{aligned}
$$

The fundamental curvatures are then given by (1.24). The covariant derivatives of a function $F: M \to \mathbf{R}$ are given by

$$
\begin{aligned}
(2.5) \qquad & \mathrm{d}F = F_1\omega^1 + F_2\omega^2 + F_3\omega^3, \\
& \mathrm{d}F_1 - F_2\omega_1{}^2 - F_3\omega_1{}^3 = F_{11}\omega^1 + F_{12}\omega^2 + F_{13}\omega^3, \\
& \mathrm{d}F_2 + F_1\omega_1{}^2 - F_3\omega_2{}^3 = F_{12}\omega^1 + F_{22}\omega^2 + F_{23}\omega^3, \\
& \mathrm{d}F_3 + F_1\omega_1{}^3 + F_2\omega_2{}^3 = F_{13}\omega^1 + F_{23}\omega^2 + F_{33}\omega^3,
\end{aligned}
$$

its Laplacian by

$$(2.6) \qquad \Delta F = F_{11} + F_{22} + F_{33}.$$

Theorem 2.1. *Let $M \subset E^4$ be a hypersurface. Suppose:* (i) *∂M consists of umbilical points*; (ii) *on M,*

$$(2.7) \qquad H_1 = \text{const}$$

and

$$(2.8) \qquad 3H_1H_3 - 4H_2{}^2 + H_1{}^2H_2 \geqq 0.$$

Then M is a part of a hypersphere or a hyperplane.

Proof. From (2.7) and (2.2),

$$(2.9) \qquad \alpha + \lambda + \xi = 0, \quad \beta + \psi + \sigma = 0, \quad \gamma + \varrho + \varepsilon = 0;$$

from (2.9) and (2.3), we easily get

$$
\begin{aligned}
(2.10) \qquad & A + D + F = D - G + F - S, \quad B' + M + I = B' - B + I - T, \\
& G + P + R = G - D + R - V, \quad C' + Y + U = C' - C + Y - N, \\
& S + V + W = S - F + V - R, \quad X + Q' + Z = Q' - Q + X - E.
\end{aligned}
$$

Now, let us consider the function $f: M \to \mathbf{R}$,

$$
\begin{aligned}
(2.11) \qquad f = H_1{}^2 - 3H_2 = & \frac{1}{2}(a_{11} - a_{22})^2 + \frac{1}{2}(a_{11} - a_{33})^2 + \frac{1}{2}(a_{22} - a_{33})^2 \\
& + 3a_{12}^2 + 3a_{13}^2 + 3a_{23}^2.
\end{aligned}
$$

Of course, $f \geqq 0$ on M, and we have $f = 0$ just at the umbilical points of M. After a somewhat lenghty calculations, we get

$$
\begin{aligned}
(2.12) \qquad \frac{1}{3}\Delta f = & \alpha^2 + \psi^2 + \varepsilon^2 + 3(\beta^2 + \gamma^2 + \lambda^2 + \xi^2 + \sigma^2 + \varrho^2) + 6\theta^2 + 3H_1H_3 \\
& - 4H_2{}^2 + H_1{}^2H_2.
\end{aligned}
$$

Applying the maximum principle, we get $f = 0$ on M. QED.

Let us remark that (2.8) is satisfied on a hypersurface with non-negative principal curvatures; indeed, we have

$$
\begin{aligned}
(2.13) \qquad 3H_1H_3 - 4H_2{}^2 + H_1{}^2H_2 = & k_1k_2(k_1 - k_2)^2 + k_1k_3(k_1 - k_3)^2 \\
& + k_2k_3(k_2 - k_3)^2.
\end{aligned}
$$

Theorem 2.2. *Let* $M \subset E^4$ *be a hypersurface. Suppose:* (i) ∂M *consists of umbilical points*; (ii) M *has positive principal curvatures and an orthogonal system of lines of curvature*; (iii) *on* M,

$$(2.14) \qquad H_2 = \text{const}.$$

Then M *is a part of a hypersphere.*

Proof. The proof is straightforward. From (2.14), we find relations between $\alpha, \ldots, \varepsilon$ as well as between $A, \ldots, I$. Then we calculate f_i, f_{ii} — the function f being again defined by (2.11) — and verify

$$(2.15) \qquad \begin{aligned}&(k_2 + k_3) f_{11} + (k_1 + k_3) f_{22} + (k_1 + k_2) f_{33} + 2\alpha f_1 + 2\psi f_2 + 2\varepsilon f_3 \\ &= 2(k_2 + k_3)(\alpha + \lambda + \xi)^2 + 2(k_1 + k_3)(\beta + \psi + \sigma)^2 \\ &\quad + 2(k_1 + k_2)(\gamma + \varrho + \varepsilon)^2 + 2H_1[3\theta^2 + \beta^2 + \sigma^2 + \lambda^2 + \xi^2 + \gamma^2 + \varrho^2 \\ &\quad + 2(\alpha^2 + \psi^2 + \varepsilon^2) + (\lambda - \xi)^2 + (\beta - \sigma)^2 + (\gamma - \varrho)^2 \\ &\quad + k_1k_2(k_1 - k_2)^2 + k_2k_3(k_2 - k_3)^2 + k_1k_3(k_1 - k_3)^2].\end{aligned}$$

The only difficulty (up to the calculations) is in choosing the appropriate function f and the left-hand side of (2.15): these are to be quessed. QED.

Let us present another version of the preceding theorem.

Theorem 2.3. *Let* $M \subset E^4$ *be a hypersurface. Suppose:* (i) M *has positive principal curvatures and an orthogonal system of lines of curvature*; (ii) H_1 *attains its maximum on* M; (iii) *on* M,

$$(2.16) \qquad H_2 = \text{const}.$$

Then $H_1 = \text{const}$ *on* M.

Proof. Using the consequences of (2.16), it is easy to verify

$$(2.17) \qquad \begin{aligned}&(k_2 + k_3)(H_1)_{11} + (k_1 + k_3)(H_1)_{22} + (k_1 + k_2)(H_1)_{33} \\ &+ 2(\lambda + \xi)(H_1)_1 + 2(\beta + \sigma)(H_1)_2 + 2(\gamma + \varrho)(H_1)_3 \\ &= (\lambda + \xi)^2 + (\beta + \sigma)^2 + (\gamma + \varrho)^2 + 6\theta^2 + 3(\lambda^2 + \xi^2 + \beta^2 + \sigma^2 + \gamma^2 + \varrho^2) \\ &\quad + k_1k_2(k_1 - k_2)^2 + k_1k_3(k_1 - k_3)^2 + k_2k_3(k_2 - k_3)^2,\end{aligned}$$

$(H_1)_i$ and $(H_1)_{ij}$ being the covariant derivatives of H_1. Applying the Bochner's lemma, we are done. QED.

Theorem 2.4. *Let* $M \subset E^4$ *be a hypersurface. Suppose:* (i) ∂M *consists of umbilical points*; (ii) M *has positive principal curvatures and an orthogonal system of lines of curvature*; (iii) *on* M,

$$(2.18) \qquad H_3 = \text{const}.$$

Then M *is a part of a hypersphere.*

Proof. Let us consider the function $g: M \to \mathbf{R}$,

$$(2.19)\quad g = \frac{1}{27} H_1^3 - H_3$$
$$= \frac{1}{27}\left\{\left(\frac{1}{2} H_1 + 3k_1\right)(k_2 - k_3)^2 + \left(\frac{1}{2} H_1 + 3k_2\right)(k_1 - k_3)^2 + \left(\frac{1}{2} H_1 + 3k_3\right)(k_1 - k_2)^2\right\}.$$

The positiveness of the principal curvatures of M implies $g \geqq 0$, g vanishing just at the umbilical points. The condition (2.18) implies then

$$(2.20)\quad 9(k_2k_3g_{11} + k_1k_3g_{22} + k_1k_2g_{33})$$
$$= 2k_2k_3H_1(\alpha + \lambda + \xi)^2 + 2k_1k_3H_1(\beta + \psi + \sigma)^2 + 2k_1k_2H_1(\gamma + \varrho + \varepsilon)^2$$
$$+ 2H_1^2[k_1(\varrho^2 + \sigma^2) + k_2(\gamma^2 + \xi^2) + k_3(\beta^2 + \lambda^2)] + 2H_1^3\theta^2$$
$$+ H_1^2[k_1k_2k_3^{-1}(\xi^2 + \sigma^2) + k_1k_3k_2^{-1}(\lambda^2 + \varrho^2) + k_2k_3k_1^{-1}(\beta^2 + \gamma^2)]$$
$$+ H_1^2[k_1(k_2^{1/2}k_3^{-1/2}\xi + k_3^{1/2}k_2^{-1/2}\lambda)^2 + k_2(k_3^{1/2}k_1^{-1/2}\beta + k_1^{1/2}k_3^{-1/2}\sigma)^2$$
$$+ k_3(k_2^{1/2}k_1^{-1/2}\gamma + k_1^{1/2}k_2^{-1/2}\varrho)^2]$$
$$+ H_1^2H_3[(k_1 - k_2)^2 + (k_1 - k_3)^2 + (k_2 - k_3)^2],$$

and the maximum principle implies the assertion. QED.

The corresponding result is contained in the following

Theorem 2.5. *Let $M \subset E^4$ be a hypersurface. Suppose:* (i) *M has positive principal curvatures and an orthogonal system of lines of curvatures*; (ii) *H_2 attains its maximum in M*; (iii) *on M,*

$$(2.21)\quad H_3 = \text{const}.$$

Then H_2 = const *on M.*

Proof. The differential consequences of (2.21) imply

$$(2.22)\quad k_2k_3(H_2)_{11} + k_1k_3(H_2)_{22} + k_1k_2(H_2)_{33}$$
$$= 2H_2\theta^2 + H_3[k_3(k_1 - k_2)^2 + k_2(k_1 - k_3)^2 + k_1(k_2 - k_3)^2]$$
$$+ (k_1\varrho + k_2\gamma)^2 + (k_2\xi + k_3\lambda)^2 + (k_1\sigma + k_3\beta)^2$$
$$+ k_1^2(1 + 2k_2k_3^{-1})\,\sigma^2 + k_1^2(1 + 2k_3k_2^{-1})\,\varrho^2 + k_2^2(1 + 2k_1k_3^{-1})\,\xi^2$$
$$+ k_2^2(1 + 2k_3k_1^{-1})\,\gamma^2 + k_3^2(1 + 2k_1k_2^{-1})\,\lambda^2 + k_3^2(1 + 2k_2k_1^{-1})\,\beta^2,$$

and we are done. QED.

The reader is invited to produce more deep theorems. He has to overwin "just" three difficulties: 1) to choose a convenable function denoted here by f or g resp.; 2) to choose the appropriate left-hand side operators as it has been done in (2.12), (2.15) or (2.20) resp.; 3) to choose the conditions on H_i in such a way the right-hand sides of (2.12) etc. be non-negative. Unfortunately, I see no systematic way to accomplish this.

At the end of this section, let us sketch the proof of a very general theorem. The main open problem is the weakening of its suppositions.

Theorem 2.6. *Let $M \subset E^4$ be a hypersurface. Suppose:* (i) *∂M consists of umbilical points;* (ii) *M has an orthogonal system of lines of curvature;* (iii) *there is a function $F(x, y, z)$ defined for positive x, y, z such that*

$$(2.23)\qquad \frac{\partial F}{\partial x} \geqq 0, \quad \frac{\partial F}{\partial y} \geqq 0, \quad \frac{\partial F}{\partial z} \geqq 0, \quad \left(\frac{\partial F}{\partial x}\right)^2 + \left(\frac{\partial F}{\partial y}\right)^2 + \left(\frac{\partial F}{\partial z}\right)^2 > 0$$

and, on M,

$$(2.24)\qquad F(H_1, H_2, H_3) = 0;$$

(iv) *on M, we have (for $i, j = 1, 2, 3$)*

$$(2.25)\qquad k_i > 0,$$

$$(2.26)\qquad k_i \leqq \frac{1}{2} H_1 \quad \text{if} \quad \frac{\partial F}{\partial H_2} > 0,$$

$$(2.27)\qquad \sqrt{5} - 1 \leqq 2k_j^{-1}k_i \leqq \sqrt{5} + 1 \quad \text{if} \quad \frac{\partial F}{\partial H_3} > 0,$$

$$(2.28)\qquad \sum_{i,j,l} (k_i - k_j)\,[(H_l)_{ii} - (H_l)_{jj}]\,\frac{\partial F}{\partial H_l} \geqq 0.$$

Then M is a part of a hypersphere.

Proof. The condition (2.24) implies

$$(2.29)\qquad \left[\frac{\partial F}{\partial H_1} + (k_2 + k_3)\frac{\partial F}{\partial H_2} + k_2 k_3 \frac{\partial F}{\partial H_3}\right] f_{11}$$
$$+ \left[\frac{\partial F}{\partial H_1} + (k_1 + k_3)\frac{\partial F}{\partial H_2} + k_1 k_3 \frac{\partial F}{\partial H_3}\right] f_{22}$$
$$+ \left[\frac{\partial F}{\partial H_1} + (k_1 + k_2)\frac{\partial F}{\partial H_2} + k_1 k_2 \frac{\partial F}{\partial H_3}\right] f_{33}$$
$$= \sum_{i<j} (k_i - k_j) \sum_l [(H_l)_{ii} - (H_l)_{jj}]\,\frac{\partial F}{\partial H_l}$$
$$+ \sum_{i<j} k_i k_j (k_i - k_j)^2 \left(3\frac{\partial F}{\partial H_1} + 2H_1 \frac{\partial F}{\partial H_2}\right)$$
$$+ H_3 \sum_{\substack{i<j \\ j \neq l \neq i}} (k_i + k_j - 2k_l)^2\, k_l \cdot \frac{\partial F}{\partial H_3} + \sum_i \Phi_i \frac{\partial F}{\partial H_i}$$

with

$$(2.30)\qquad \Phi_1 = (\alpha - \lambda)^2 + (\alpha - \xi)^2 + (\lambda - \xi)^2 + (\beta - \psi)^2 + (\psi - \sigma)^2 + (\beta - \sigma)^2$$
$$+ (\gamma - \varrho)^2 + (\gamma - \xi)^2 + (\varrho - \varepsilon)^2$$
$$+ 6(\beta^2 + \gamma^2 + \lambda^2 + \varrho^2 + \xi^2 + \sigma^2 + 3\theta^2),$$

$$\begin{aligned}
\Phi_2 = {} & 2(k_1 + 3k_2 + 3k_3)(\lambda^2 + \xi^2) + 2(k_2 + 3k_1 + 3k_3)(\beta^2 + \sigma^2) \\
& + 2(k_3 + 3k_1 + 3k_2)(\gamma^2 + \varrho^2) \\
& + 2k_1(\alpha - \lambda - \xi)^2 + 2k_2(\psi - \beta - \sigma)^2 + 2k_3(\varepsilon - \varrho - \gamma)^2 \\
& + 2(k_2 + k_3 - k_1)\,\alpha^2 + 2(k_1 + k_3 - k_2)\,\psi^2 \\
& + 2(k_1 + k_2 - k_3)\,\varepsilon^2 + 12H_1\theta^2, \\
\Phi_3 = {} & K_{123}(\lambda - \xi)^2 + K_{213}(\beta - \sigma)^2 + K_{312}(\gamma - \varrho)^2 \\
& + k_2k_3\{(\alpha + K'_{231}\lambda)^2 + (\alpha + K'_{321}\xi)^2\} \\
& + k_1k_3\{(\psi + K'_{132}\beta)^2 + (\psi + K'_{312}\sigma)^2\} \\
& + k_1k_2\{(\varepsilon + K'_{123}\gamma)^2 + (\varepsilon + K'_{213}\varrho)^2\} \\
& + (k_1k_2 + L_{123})\,\lambda^2 + (k_1k_2 + L_{213})\,\beta^2 + (k_1k_3 + L_{132})\,\xi^2 \\
& + (k_1k_3 + L_{312})\,\gamma^2 + (k_2k_3 + L_{231})\,\sigma^2 + (k_2k_3 + L_{321})\,\varrho^2 \\
& + 2\Big(2\sum k_i^2 + \sum_{i<j} k_ik_j\Big)\,\theta^2;
\end{aligned}$$

$$K_{ijl} = \left(k_i - \frac{1}{2}k_j - \frac{1}{2}k_l\right)^2 + \frac{3}{4}(k_j - k_l)^2 + k_i^2 - (k_j - k_l)^2,$$

$$K'_{ijl} = k_i^{-1}(k_j - 2k_l),$$

$$L_{ijl} = 3k_j^{-1}k_l^3 + k_ik_l - 2(k_i - k_l)^2 + 4k_j^{-1}k_l[k_j^2 - (k_i - k_l)^2],$$

and we are done. QED.

The reader is, of course, able to enjoy himself otherwise than by checking (2.29). He even may exhibit new methods for proving results of this form.

Bibliography

[1] AFWAT, M.; ŠVEC, A.: Global differential geometry of hypersurfaces. Rozpravy ČSAV, Academia, Praha 1978.

[2] BOCHNER, S.; YANO, K.: Curvature and Betti numbers. Princeton Univ. Press, 1953.

[3] HUCK, H.; ROITZSCH, R.; SIMON, U.; VORTISCH, W.; WALDEN, R.; WEGNER, B.; WENDLAND, W.: Beweismethoden der Differentialgeometrie im Großen. Springer-Verlag, Berlin—Heidelberg—New York 1973.

[4] VEKUA, I. N.: Verallgemeinerte analytische Funktionen. Akademie-Verlag, Berlin 1963 (translation from the Russian).

See also:

[5] SULANKE, R.; WINTGEN, P.: Differentialgeometrie und Faserbündel, VEB Deutscher Verlag der Wissenschaften, Berlin/Birkhäuser Verlag, Basel—Stuttgart 1972.

[6] ŠVEC, A.: Contributions to the global differential geometry of surfaces, Rozprawy, ČSAV, Academia, Praha 1977.